HEARING SILENCE

First published in Great Britain by Black Apollo Press, 2008
Copyright © Laurinda Brown and Alf Coles 2008

The moral right of the authors has been asserted.

A CIP catalogue record of this book is available at the British Library.

ISBN: 9781900355599

Design: David Cutting, DCG DESIGN, Cambridge

HEARING SILENCE

LEARNING TO TEACH MATHEMATICS

Laurinda Brown & Alf Coles

In order to listen, it is necessary
to keep silence ...

(Father Gratry, 1944, p518)

I need to make works that anticipate, but do
not attempt to predict or control, the future.
In order to understand time, I must work
with the past, the present and future.

(Goldsworthy, 2000, p7)

To Dick Tahta in memory

The image on the front cover was taken at a branch meeting of the Association of Teachers of Mathematics (ATM) many years ago, and reproduced here with kind permission of the ATM. Dick Tahta used the image in an article he wrote for the 'How I Teach' issue of *Mathematics Teaching* 139, the journal of the ATM: 'In order to listen it is necessary to keep quiet. Like the nun in this picture, I too try to still my lips with my fingers.' (p. 21)

CONTENTS

PREFACE

This book is about a collaboration between two mathematics teachers. It offers a historical narrative of their joint research into what is involved in becoming a teacher of mathematics. This is a story of change and the process of change.

Their work began in 1995. Alf had recently taken up his first school-teaching post, and Laurinda was a university lecturer who was interested in observing and interpreting a new teacher's developing practice. But this soon shifted into a joint exploration of ways of teaching mathematics in Alf's classroom, Laurinda being an intermittent visitor of at most once a week. They planned lessons together, and wrote up their observations on each other's teaching at the time.

Laurinda would interview Alf every September, shortly after his first few lessons of the new school year. Transcripts of two such interviews, the very first one in 1995 and another four years later, can be read in full in an Appendix. These indicate something of the journey that Alf had made in his teaching. It seemed appropriate to offer an account of this development.

The following six chapters focus on some specific aspect of this collaborative research. In each case, a preliminary discussion of the issues involved is presented in the form of a dialogue. Relevant extracts from the transcripts of the above-mentioned interviews are quoted here.

The opening dialogue is then followed by reflections from Alf on his professional development as a teacher. A third section provides some further comment taken from reports presented by the authors to an annual international conference of mathematics educators.

A final section provides an account of an actual lesson or lesson sequence that illustrates the theme of each chapter. This section also serves to document an exploration of ways of presenting lesson 'write-ups' that can support teachers – going beyond simply offering a script to follow.

1 DEVELOPING TEACHING

1.1 Agreement to work

Laurinda: You had come to a meeting with Charlie, your PGCE tutor, and had been telling me stories from your classroom over lunch. When I said I could come to London for a day and work with you in your school I remember that you looked surprised.

Alf: I could be articulate in meetings with adults about my ideas for mathematics teaching but I knew there was a gap between these ideas and the reality of my classroom. But at the same time I think I must have known that if I was going to work on putting my ideals into practice, I needed some help.

Laurinda: Yes, you are articulate. Long words to year 7s who shrugged their shoulders and looked blank. It was as if you looked down on them or couldn't get alongside them or be curious about where they were. Learning to teach is a journey of self-discovery, quite painful at times.

Alf: I've always been more comfortable with people older than myself. You weren't articulate about what you did when we met, but I recognised that you could 'get alongside' my students. I think part of me wanted answers when I started teaching.

Laurinda: I had been researching my practice as a teacher educator some years before meeting you. I knew what I did but did not have a vocabulary to describe it. This was a strong motivation to work with a mathematics teacher in their early years of teaching who wasn't finding things easy and who hadn't gone through the Post Graduate Certificate of Education (PGCE) course on which I worked. I did have some ethical concerns though about working with someone who was so articulate when one saying is that 'the good practitioner is inarticulate about their practice'. Would getting you to talk on tape and write about your practice get in the way of your learning to teach?

Alf: I was used, after taking philosophy courses at university, to discussing and analysing ideas. I'm sure that's partly why I felt articulate in meetings – I could play that game, and I'm not sure how much it helped my classroom teaching. One thing you have always been strong about is that, when we talk about lessons, we stay with the detail of reconstructing what happened for as long as possible before moving to try and interpret or analyse events. The talking always felt as though it took me deeper into my practice, literally allowing me to see more detail, rather than provoking any analytical or philosophical distancing.

Laurinda: 'Staying with the story' still seems an important part of our work together, looking at the details of practice. You said at the time, when observing me teach your year 8 on the second visit to the school in London, that you recognised something that you valued about the role of the teacher.

Alf: I'd read an article, given to me by Charlie, about the role of the teacher (see *Theory Box 1.1* opposite) which had certainly fed into my image of how I would like to be in a classroom. But I'm not sure that the article gave me any strategies for how to get there. Watching you teach my year 8 class I recognised a lot of what I had read and I wrote up the lesson soon afterwards (see Section 1.4). There's something here about the melding of theory and practice that wasn't natural for me to do at the time.

Laurinda: In 1995 your ideal was to emulate the teacher in the article but you had no image of how that looked in practice. When, in an interview, you first engaged with the process of re-telling a lesson you started to justify and interpret the events. In 1999 you were used to being interviewed about your first lesson with a new class of students and there are very long blocks of text from you as you try to, what in 1995 you described as, 'stay with the story' of the lesson.

Alf: You insisted then that I described my account of the details of what was happening rather than 'accounting for' (Mason, 1996, pp. 23-24) the events. This is a discipline I've found useful throughout the six years of our work together, and we have continued to focus on what has been done, rather than on our justifications for our actions.

Laurinda: Within the transcripts there is a contrast between 1995 and 1999 in your descriptions of classroom management issues. Initially you didn't have much conviction about what you were doing either in setting up expectations of behaviours or discipline.

Theory Box 1.1:
From 'The role of the teacher' (Wheeler, 1970, p. 27)

If we watch this teacher at work we see that:

- he teaches the whole class or a group of the whole class much of the time
- he sets the situation, giving essential information, but beyond that tells the children nothing
- he obtains as much feedback from the children as possible, by observing, asking questions, and asking for particular actions
- he works with this feedback immediately
- he never collects answers from the whole group to check that they all have the correct one, nor does he collect all the different answers they may have obtained without working on their differences
- except on rare occasions he does not indicate whether a response is right or not, though he often asks children which it is
- he accepts errors as important feedback telling him more than correct responses, and by directing the children's attention back to the problem he urges them to use what they know to correct themselves
- he does not praise or blame particular responses or particular children, though he may exhort or reassure by expressing his faith in the children's capabilities
- his dialogues with the children contain phrases like:
    Look at the problem
    Look at what you have done
    Listen to what you are saying
    Don't guess: tell me what you know
    You know you can do this: try it again
    Is it right?
    How do you know that?
    Are you sure?
- he respects a child's right to opt out, and if the feedback from the children assures him he is trying to make them go too fast or too far he abandons his attack at once
- if he sees that a particular child in a group who is not immediately succeeding can be challenged so that he conquers his hesitations, he gives all his attention to that child at once
- if he sees that a particular child in a group is working on the problem but not contributing to the group, he leaves the child alone until he wishes to take part
- he appears to an uninvolved observer to be impersonal in his approach, giving no favours to anyone and taking success for granted, but showing no disappointment if it does not arrive.

Watching him, we are conscious that everything he does directs the children to attend to the problem and to their own actions in tackling it.

Alf: I am amazed that I was able to walk into a classroom then given how unsure I was about what classroom rules I wanted:

> It's something I don't feel I've resolved anyway, the rules in general. That's where the doubt comes in in the first lesson, that I'm not really sure overall exactly what I want to expect. (Appendix 2, 1995, 2.26)

Laurinda: This contrasts markedly with the detail in 1999 in just a few minutes of meeting and greeting:

> ... then I was just in the corridor and some people were coming up wondering where C11 was, so I told them to line up outside the door. They had obviously been to different lessons because they were coming in in dribs and drabs. So, about eight of them had lined up and so I got three boys to go in and sit down, and I asked them to go and sit, the first three people to go in and sit at a table. A few more people started arriving and I got the next four or so to go and sit at a table. (Appendix 1, 1999, 1.2)

Alf: In 1995 the rules which I shared with my class were departmental ones:

> Basically I was walking in not quite sure exactly what I wanted to do with them. I came in. We have these rules for mathematics lessons that I've never given to a class. We're meant to do that first of all. So I decided that I would with this class. And that actually felt alright 'cos it just got them quietly working on something [copying rules into their books]. (Appendix 2, 1995, 2.4)

Laurinda: These were certainly not rules which you saw as important. Again, in contrast, in 1999:

> ... and then I said 'something I haven't mentioned is about the rules of the classroom but I'm sure lots of teachers have been talking about that' and some girl mouthed 'all of them'. So I said that they'd been in schools for 6 years and they knew what sensible rules of the classroom were about and I was sure they were mature enough to understand what was expected of them and it would become apparent if they were doing things I wasn't happy with and that one thing I had noticed was that some people were calling out without their hands up and that was a reference back to the beginning of the lesson and so I said that that was one thing, because if we were going to have discussions that was going to make it very difficult. So, that was one thing that I didn't want them to do. (Appendix 1, 1999, 1.16)

Alf: And at the end of that lesson I phoned a parent of one of the students:

> So that was it and I let them out table by table. Then one person stayed and two people were being quite silly, so I spoke to them even though it was the end of the lesson. And one boy when I asked them to get their homework diaries out said 'I haven't got mine' and I could see it in his bag. So, I told him to get it out and I actually phoned his mum and said 'Very surprised in the first lesson, something I can expect hardened year 9s to do when they don't want to write down a homework'. She said she'd have a word with him. (Appendix 1, 1999, 1.50)

Laurinda: There were issues to do with discipline in the early days of our work together and you spent a lot of your energy on controlling the class. In our agreement to work together was a joint belief in the power of engaging students in mathematical activities to reduce problems in classroom management. The necessary routines for managing a class are not the prime focus of your attention any more, which they were at times in 1995. You now act when necessary. It is as if you can develop your mathematics teaching because you need to spend less of your conscious energy on controlling the class. I find it is striking that often in the comparison of the two transcripts the roots and seeds of the 1999 Alf are there in the 1995 one.

1.2 Starting to teach (Alf)

I began teaching in England in 1994. During my PGCE year I had wanted to establish a humane classroom in which students expressed their autonomy and were engaged in self-motivated and creative activity. I had been influenced by ideas such as those in **How Children Fail** (Holt, 1990) and wrote the following in 1995:

> July spelt the end of my first year of teaching. In the course of the year, I gained a sharpened image of how I would like to be in class, and of the ways in which I would like my students to be working. Along with this clarification, however, came a widening gap between the image and what actually was happening. Ideally, I see myself being matter-of-fact, avoiding condescension and blame and even praise; capturing attention and emotion and directing this into students' own awarenesses. In reality, I rarely held the classes' attention. I realise I did condescend to my year 7s and often became negative.

> I would like my students to feel that Maths was something that connected with them, that its roots lay in their own intuitions, and that it can be tackled independently. I hoped the study of Maths would help students to realise the potential and power of their minds. None of this happened. Indeed, many students now lack the faith that there is much to be gained from Maths at all. Far from opening minds, it is clear that some students have a diminished sense of their capabilities and an impoverished self-esteem. The fundamental features of what I envisaged in my classroom have not materialised. (Diary, Alf, 07/95)

I found it difficult to use these realisations positively. They merely engendered despair. My beliefs about what I wanted in a classroom were not seemingly linked to any actions that might help realise them. In reflection now, I see a consistency with my University days. Most of the philosophy I studied there was similarly separated from what I did in my life. I could write analytic essays about, for example, Aristotle's **Ethics** (1980), without that having any impact on how I thought about the ethical decisions in my own life. I am not sure I even recognised that I made any ethical decisions. Before the end of the summer term, in my first year of teaching, I met Laurinda who encouraged me to look back on the year and just think of times I approached my ideal image of how I would like to be in class. I wrote at the time about this incident:

> Looked at in this way, worries about whether lessons had been a 'success' or whether anything had been 'learnt', disappeared. I was able to focus on how I had used myself. The first memories that came were of classes where the students also worked in a way that approached my image. These were the high points of the year. However, after that came lessons or half lessons, which were often far from effective, where at least I felt I was on the right lines. There was scope in what I was doing to be as I wanted. These were often half-forgotten moments, filed away as unsuccessful experiments, but I think that they may be the occasions on which to build. (Diary, Alf, 07/95)

Something about the offer to stay with giving detailed descriptions of particular incidents allowed me to avoid the debilitating judgements that previously had coloured whole lessons and indeed the whole of my first year of teaching. I was able to identify things I had done which seemed connected with times when students were working in ways that I valued. Two stories initially came to mind (I talk about these in detail in *Section 2.2*) and I recognised that a sameness between these stories was my own, deliberate, silence. What seemed important at the time for me in developing my teaching was that I had something in mind which I could focus on doing (*e.g.* remaining silent during the start of the lesson) and which could inform my planning.

During my second year of teaching, Laurinda spent two days each term in school, and we worked together in my classes. I wrote after her first visit:

> Discussing the lessons afterwards Laurinda offered observations on a level that I was unused to. ... There were many things going on in those classes which were exactly the type of thing I would say I wanted – yet what became clear was that I was not aware of when they actually occurred. Laurinda's comments however did have a ring of truth for me – on some level I must have been aware of these things. (Diary, Alf, 10/95)

What I believe this diary entry captures is the start of a shift in my attention from judgmental perceptions about the whole lesson (*e.g.* 'this is/is not going well') to a more detailed awareness of what is happening for individuals. I wrote at the time:

> One discipline that has also come out of the work with Laurinda is that of 'staying with the story'. In my notes on teaching in the first year the observations are in general distant, about whole classes, with observation and analysis all mixed in. What I have been working on this year is forcing myself to hold back the analysis and stay with stories about individuals or groups of individuals. Analysis (or synthesis) from this data then has the possibility of throwing up something I had not been aware of before. There was something previously about the mixing of analysis and observations that meant that I was never surprised – everything that happened was seen through the filter of what I expected. There was little chance of my accessing those things I did that I was unaware of – but which yet had profound effects. (Diary, Alf, 10/95)

I experienced various realisations from looking back over the detailed observations that I would sometimes write after lessons. An early realisation was a sameness about comfortable moments in class, that was to do with me not really being needed by the students. Such awarenesses could then become the focus for planning – *e.g.* 'how could I structure this lesson so that I will not be needed by the students?'. By this phrase I meant that I did not want to be needed by the students to tell them if what they were doing was right/wrong or to tell them what they had to do next. I was most comfortable if the students had a question or problem which they were discussing amongst themselves, trying to convince each other about what was the case, without the need for me to authorise or validate their thinking. Again, from my diary:

> My own observations about myself are, perhaps, of little interest. I feel however that I have some mechanism for reflecting on my practice that helps to deepen my awareness of what I am doing – both after the classes and in the moment.

> I have experienced, this year, classes working in an autonomous and motivated way, coming to important realisations over a series of lessons. This is by no means the norm but I feel it could be. I have become aware of my awareness of what energy students have. When six students came up and showed me their homework before the beginning of a class I was able to see this not just as a distraction to what I was wanting to do but as an indication that there is perhaps something here that we can all work on. (Diary, Alf,10/95)

I had begun the journey of being able to reflect on and develop my practice.

1.3 On teaching strategies (Laurinda) – see Brown (1995)

What did I mean by purposes? What did I mean by teaching strategies?

I started teaching mathematics in September, 1973 and worked in the same school later as second in department and then, until 1987, as Head of Department. During that time I was not aware of the variety of different methods of teaching mathematics because the department had a strong philosophy based on interactive whole class teaching and I did not visit other schools. I assumed, just did not think about it, that other schools were the same. On leaving this school to take up a one year secondment as Mathematics Editor at the Resources for Learning Development Unit (RLDU), a curriculum development post, I was invited to work with other departments and visit schools and classrooms. At first, I was in shock about the sheer variety of practice, and developed a theory that no two departments were in any way similar – a theory which still holds, I might say, independent of the various central changes in curriculum policy.

At that stage of my career, in 1988, I was finishing off a taught Masters in Mathematics Education at the University of Bristol with a dissertation to complete. I decided to continue to look at classrooms and asked the mathematics advisory teachers to point me in the direction of some teachers who they would consider to be particularly good in teaching using a range of structures from individualised learning to whole class teaching. The dissertation was titled 'The influence of teachers on children's image of mathematics' (Brown, 1988, 1995) and attempted to probe the links between what the teachers said they were trying to do, their philosophies of mathematics and mathematics teaching, and what their students' impressions of learning mathematics were. One class from each of 4 teachers, A, B, C and D and 6 students from that class were the focus of the research.

I could not find within these classrooms students with a stereotypical view of mathematics as something alien and difficult. It was challenging and enjoyable, satisfying when you knew something you did not know before. To share just one example:

> I prefer having to solve it myself. It gives you that satisfaction of not having to take it from a book. I enjoy mathematics. I find it more of a challenge than a chore. (Student C2, Brown, 1988, p. 84)

What were these teachers, all very different, doing to have these effects? Standardisation was not the key! In fact I wondered if, having observed each other teach, what they would have taken away? It was difficult to imagine them simply trying to become the other person.

One incident stood out for me. Teacher C used a particular, what I called, teaching strategy, which was obviously well-rehearsed, with a particular class:

> ... a pupil offered an explanation of how they had begun to tackle a problem. The other pupils were invited to close their eyes and put up their hand if they had started in the same way. In fact, in the lesson observed, only two pupils did so. An alternative start was requested and the pupils again closed their eyes and put up their hands if this was their way of starting. The process continued with more information being collected and these different starts were then used for further exploration (as the teacher asked): What was the aim of the people who drew the radius? (Brown, 1995, p.152)

I knew immediately that I would use this teaching strategy if I was teaching again. In conversation with Teacher C afterwards it transpired that he had, in his turn, recognised the teaching strategy as a useful one for him through observing a primary school teacher. He was clear that it did not work with all the classes that he taught, but that where it became established as a routine it helped him to establish the classroom ethos that he valued.

In trying to find a way of thinking about what teachers actually do to achieve the atmosphere in their classrooms which they valued, I interviewed teachers about the first mathematics lessons that they had given at the start of a school year to classes which they had not taught before. What were the links between such stories and how the teacher operated in classes that they had taught for more than one year? I started using a confusing variety of words such as 'strategy', 'tactic', 'skill' and 'technique' to describe what I saw. These terms, as well as the commonly used 'micro-behaviour', 'micro-strategy' and 'macro-strategy', were referred to by different authors without consistency. Sometimes the teacher used a specific and repeatable behaviour over and over again with marked effect and at other times what was repeated seemed more nebulous and global such as 'asking questions'. How could I find a way of describing what I was looking for?

I became aware through the work at the RLDU that there was a gap between how teachers described what they did which was tied into their beliefs about mathematics, such things as Alf's 'autonomy', and their actual practice or teaching strategies. When I went and observed teachers in the classroom who were part of my working groups I started to use the word 'purpose' without really having thought why or questioned what I meant by the word (for links to the literature related to 'purposes' see Brown and Coles, 2000). I needed a word to describe the overlapping themes in the way in which the teachers talked about what they were doing. These themes were not at the level of their beliefs nor at the level of their behaviours or teaching strategies but in a middle position closely related to their actions. For example:

Purpose 1: Knowing what they know

At the start of a topic or theme how can you find out what the individual students in your class know and where they find problems so that you can plan for progression?

Teaching strategies:

− invite the students to make posters or write in response to 'Tell me what you know about ...'
− diagnostic 'can do' sheets
− open-ended starter

e.g. You're going to be working on area and you invite the students to draw shapes with area 8. This can be constrained by using square dotty paper and inviting the corners of the shapes to be on the points of the grid.

(From notes written as part of the Classroom Methodology group for SMP 'New Roots' (now 'Interact') early development meeting, Laurinda, 1994)

What seemed important here is that, after I had identified a common theme with a label such as 'knowing what they know', the focus on that purpose in discussion with the other teachers allowed them to describe the detail of other teaching strategies they used to achieve that purpose. There could, in fact, be very many teaching strategies related to a particular purpose that an experienced teacher would be choosing from dependent on the complex factors of the particular students in, for instance, their particular classes, resources and atmosphere.

My work interviewing teachers after their first lessons at the start of the school year had sensitised my own powers of noticing purposes when other teachers talked about their classrooms. On starting to work with PGCE students, I was concerned to find ways of provoking the kind of learning experience that I had with Teacher C, where I recognised a teaching strategy that I would use for a particular purpose.

Giving my PGCE students 'tips for teachers' from my own experience was an impossible task. No two situations were ever the same and so any particular teaching strategy could always be wrong and almost certainly would be in some cases (*e.g.* sending a student outside the room conflicting with school policy in one environment and not in another). Similarly, but differently, engaging in long philosophical discussions about education or what mathematics is, entrenched previously held tenets without linking them to purposes or teaching strategies.

The way of working with my PGCE students that emerged over time was to set aside a time each week, when they are in the university, for joint reflection on their teaching practices. The discipline in these sessions is to move through consideration of their teaching experiences, via the formulation of issues, to the collection of possible actions or teaching strategies (adapted from Jaworski, 1991). These university sessions begin by the students being asked to reflect back over their time spent teaching in school to identify incidents which are still with them. One student is asked or volunteers to tell their story briefly, as an 'account of' to the rest of the group and, initially, responses might take the form of questions for clarification. Stories (not necessarily the one prepared) are then told by other students that seem to be related in some way that was sparked off either through resonance ('that reminds me of . . .', 'that's similar to . . .') or dissonance, a strong feeling of difference. The group then works to try to identify any general strands or themes underpinning the stories. In this way, what I called 'purposes' emerged. These are expressed vividly in the words of members of the group and understood by them. By working at sharing teaching strategies linked to such purposes, possibilities for action in the classroom are extended. Another student then offers their original story and the process begins again. There are also times where I would have a related story suggesting an issue which hasn't been raised so far and, in this way, purposes can be offered to the group. Sometimes, these created statements remain live within the group and sometimes they do not get mentioned again and so were not found to be useful in practice. Each year the language of the PGCE group on this level of purposes is different.

Purposes are easy to articulate *e.g.* 'knowing what they know' and can therefore be easily shared by the PGCE students. They could hold very different core beliefs and still all work on common purposes to collect a range of teaching strategies. They were therefore exploring a range of possible teaching strategies having recognised the need for a particular purpose. This had become a way of working for me with groups of PGCE students and what seemed to happen is that staying with the detail of accounts of their practice led to patterns being recognised which in turn led to statements of the issue involved that I recognised as a purpose. These purposes motivated the PGCE students' actions in a classroom in a way in which their beliefs could not do. What was also fascinating was that working in this way also allowed a reworking of their beliefs about mathematics and mathematics teaching that they had brought with them to the course and which were not easily assimilated into their everyday practice as a novice teacher.

It was with these experiences that I came to the work with Alf. Would these theories, which at the time were implicit and part of my practice, work with someone who found teaching difficult who was not part of a group of PGCE students undergoing an intense learning experience?

1.4 Lesson account

On an early visit to London Alf asked Laurinda to teach his year 8 (age 12-13) class. In developing his practice, it seemed crucial, reflecting back over working together, that Alf had seen an example of someone teaching in a way that fitted with his reading and ideals. In this case, the image was strong because the students were in one of his own classes, so there was a contrast between his perceptions of how they operated in the two situations. Alf wrote at the time that watching someone else teach a group of his students and seeing that here and now, in the same classroom, these students could listen, hear and respond, both to each other and to the teacher, was an important experience for him. They did not normally behave in this way, so what was going on here? What was Laurinda doing?

The lesson was planned using a purpose of 'sharing responses': Laurinda was to use as many teaching strategies related to that purpose as she could to promote discussion in Alf's classroom. Alf saw Laurinda commenting to students about their mathematical behaviours. This sort of comment, outside of the flow of the mathematical content but related to mathematical behaviours, we have come to call a 'metacomment'. In such comments, Laurinda was focusing students' attention on their awareness of pattern.

The text below (see 'The role of teacher' overpage) was expanded from observation notes Alf took at the time of Laurinda teaching his year 8 class. Subsequently in thinking about 'The role of the teacher' article (see *Theory Box 1.1* p. 9) Alf was struck by the fact that he felt he could 'tick' Laurinda in almost all the listed attributes in this lesson. Alf added to the text, in italics, items from the list where appropriate (and changed the gender from that in the article).

At the end of the previous lesson the students had been grappling unsuccessfully with finding the rule for a quadratic function in the context of the function game.

Function Game

Without saying anything, write ...

This is the blue chalk	**2**	→	**4**	This is the red chalk
	3	→	**9**	
	4	→		

and hold up the red chalk inviting a child to fill in the gap. You now have two problems:

  a) how to indicate "rightness" or "wrongness" of the child's guess in relation to your rule
  b) how to get the chalk moved efficiently.

Although it's not always appropriate, there's a beautiful atmosphere if the lesson progresses with no one ever speaking, which can happen quite naturally and focuses everyone's attention.

A possible game with the teacher keeping the blue chalk - kids passing the red chalk amongst themselves.

2	→	**4**
3	→	**9**
4	→	**14** ☹ **16** ☺
10	→	**100** ☺
What's the rule?	→	**Times by itself** ☺

Invent your own algebraic notation or introduce standard notation.

Figure 1.1: Function Game (from Brown and Waddingham, 1982, p. 42)

Alf had set a homework to find the rules for some linear functions and to have a go at finding the rule for the function:

1	→	6
2	→	12
3	→	20
4	→	30
5	→	42
N	→	

Figure 1.2: What is the rule?

The role of the teacher – year 8, 1995

At the start of the lesson, Laurinda asked the students to compare their homework answers with that of their neighbours before we shared what they had found with the whole class. If all students in a group were stuck they were asked to find some questions they might want to ask others.

> *she never collects answers from the whole group to check that they all have the correct one, nor does she collect all the different answers they may have obtained without working on the differences*

One group had not got anywhere with the challenge for homework and were not able or not willing to address the invitation to think about a question they might ask.

> *she respects a child's right to opt-out*

Laurinda was explicit that her focus for this lesson was not so much on the actual finding of rules but on developing strategies for finding rules.

> *she sets the situation, giving essential information, but beyond that tells the children nothing*

One group of girls had not been able to find the quadratic rule but they had noticed the differences went up by two each time and so could continue the pattern. Laurinda expressed support for this when talking to them 'This is the sort of thing you could be prepared to share.' The other question was 'How do you link this with the rules?'.

> *she does not praise or blame particular responses or particular children, though she may exhort or reassure by expressing her faith in the children's capabilities*

The homework was still written on the board from the day before. Feedback was then taken on the linear rules. The link between the coefficient of N in the formula and the difference between consecutive 'outputs' was mentioned as a strategy for finding these rules.

Saying she wanted to try this out, Laurinda started a new game with another linear function and played until there were five inputs and outputs. She then asked the students to put up their hands if they knew the rule.

(on the board was)			
	1	→	2
	2	→	7
	10	→	47
	12	→	57
	14	→	67
	N	→	

Figure 1.3: The linear rule used in the lesson

she obtains as much feedback from the children as possible, by observing, asking questions, and asking for particular actions

All except three students put up their hands. The next question was whether any student could, without writing the actual rule, write up anything that might help the three who had not got it.

she works with this feedback immediately

A girl (from the group Laurinda had encouraged above) came up and wrote a five between the two and seven in the output column. She looked to see if she could do that again, but since there were no other consecutive inputs that wasn't possible. Someone suggested she put another number in. At this stage she wrote 3 → 12 (after the 2 → 7) and put another five between the seven and twelve (indicating the difference again).

(on the board was)			
	1	→	2_5
	2	→	7_5
	3	→	12
	10	→	47
	12	→	57
	14	→	67

Figure 1.4: Showing the difference of 5

A further input number was then given as a question and although many were offering to write the answer the suggestion was that someone who had not got the rule before should have a go and then we could see if the strategies were proving useful.

except on rare occasions she does not indicate whether a response is right or not, though she often asks the children which it is

The move to the 'challenge' of finding a quadratic function, which had been so difficult for them the previous lesson, led to the sharing that had been set up at the start of the lesson. The invitation was to put up on the board any rules if they had them but with a reminder that we were interested in the strategies which could be used to find a rule. Three rules were offered $N^2 + 3N + 2$; $(N + 1)(N + 2)$; $(N + 1)^2 + N + 1$. The focus of the general strategy was again re-stated and Laurinda asked whether anyone had anything helpful to say about that.

A boy said 'Isn't $(N + 1)(N + 2)$ just the same as $2N + 3$?'. Substituting in the latter formula showed that it did not generate the function. The question remained whether all could tell how the other formulae did generate the function.

she accepts errors as important feedback telling her more than correct responses, and by directing the children's attention back to the problem she urges them to use what they know to correct themselves

There was an explicit request for them to raise their hands if they were able to tell how any or all of the other rules generated the function. The girls mentioned at the beginning had continued to work on what they had noticed about the differences throughout the lesson . . .

if she sees that a particular child in a group is working on a problem but not contributing to the group, she leaves the child alone until she wishes to take part

. . . they suggested a structured way of decomposing the rule:

$$
\begin{array}{ccccc}
1 & \rightarrow & 2 \times 3 & \rightarrow & 6 \\
2 & \rightarrow & 3 \times 4 & \rightarrow & 12 \\
\ldots & & & &
\end{array}
$$

Figure 1.5: A structured way of decomposing the rule

This structure connected with the $N \rightarrow (N + 1)(N + 2)$ expression and made the link with the rule explicit for many. An offer of another quadratic rule and their difficulties in finding it showed that something more was needed before a general strategy emerged that worked for all quadratics and this was left as the challenge for the next lesson.

she teaches the whole class or a group of the whole class much of the time

(E-mail communication from Alf to Laurinda, 31/01/96)

2 CLARIFYING AWARENESSES

2.1 Forging a language

Laurinda: When I interviewed you in 1995, I was aware that you had little sense of what you were trying to achieve in your teaching or the classroom culture you were trying to encourage. You may recall what I said at the time:

> The question there would be 'Is that something you're trying to set up?' so it's back to this 'what do you want to happen in that kind of lesson?' which might set a pattern for what might happen in other lessons. Can you remember what you told them about what they had to write? (Appendix 2, 1995, 2.47)

Alf: I remember talking about what I tried to do with one class in that interview:

> With my year 9 class who I had in year 8 one of the things that I said to them at the beginning is that 'one of my aims this year is to try to make you independent thinkers', although who knows what independent means. That felt quite nice because I could come back to that. (Appendix 2, 1995, 2.65)

Laurinda: I recognised this 'coming back to' as a comment about behaviours which supported the students' development as independent thinkers. This is what I would, in 1995, have called a metacomment (following Bateson, 1979). Now you are consciously metacommenting in your classroom. A shared language of metacommenting has developed between us.

Alf: From the first lesson of the year I now tell students that they have a purpose for the year of 'becoming a mathematician'. What I see as mathematical behaviours are to do with my past history but I then comment about the mathematical behaviours of the students as I observe them. These metacomments, related to a purpose, seem to support the students in realising how to act in my classroom.

Laurinda: And I try, as shown in the more recent interview, to support your continuing to take opportunities for metacommenting:

> This is exactly what mathematicians do, they'll extend the pattern, we'll all get used to that and expect there to be structure. There's this sense that 'how do the kids know that they're not randomly doing . . .?' They can't know that without the metacomment. (Appendix 1, 1999, 1.45)

Metacommenting is now well established in your practice.

Alf: And you noticed that the metacomments related to 'becoming a mathematician' become less present as the year goes by.

Laurinda: It's as though the behaviours related to earlier metacomments become implicit in the culture of the classroom, carried out individually by students and their teacher and also collectively. We've developed a shared language out of our practice of teaching and learning. Early on we used to tell accounts of classroom incidents and then reflect on these to raise issues. I can remember us working hard for about a year or so worrying about interpretations and what would someone else's multiple stories of a single event do for you? What was it all for? In the end we seem to have moved away from interpretations and into looking at the stories in more detail, developing new labels as we go. I'm interested in the shifting levels of communication from a collection of accounts of practice to the recognition of a purpose. Have you got any current examples for us to work on?

Alf: Here are two accounts from my teaching diary:

> Account 1: A student in my set 8 out of 8 year 11 group called me over, having got stuck on a question, started describing what his difficulty was and, without me saying anything, sorted himself out. I asked him what had just happened and he replied that all he had done was talk about where he was stuck. I suggested he didn't need me there to do this. About ten minutes later, in which time he had seemingly been very focused, he exclaimed: 'It worked!'. I asked: 'What?' and he replied: 'What you said! I got stuck on this question and talked to myself and worked out what to do!'

> Account 2: My further mathematics A-level student will regularly come to class with a list of questions and exercises he got stuck on, sit down in the (always same) chair and sort out his difficulties with very

little intervention from me. He commented for some time: 'What is it about this chair?' and then one day came in saying he had got stuck on something and imagined he was in the chair, which he had found helped him! (Diary, Alf, 1999)

Laurinda: In telling these two stories together, you must be aware of some similarity. In both you are talking about the awareness of others about their ways of working.

Alf: Yes, I would agree.

Laurinda: You now have this awareness available to focus the future mathematical behaviour of your students. Whenever you sit silently whilst a student sorts themselves out, the offer of 'What did I do?' could be made. This is feeling like a purpose. Where's the shifting level of communication?

Alf: In the first account the student is stuck trying to answer a question. My presence and attention supports him to be able to look at his own difficulties, via articulating them, and find a way through for himself. The shift in level of communication comes from the student, with my attention, being able to think *about* what he is doing. Through shifting level he learnt something about the problem he was working on. After the comment of: 'What did I do?' the student thought back on what had just happened and shifted level again in recognising that he did not need me in order to reflect (shift level) on where he was stuck. The fact that he is then able to use this strategy I'd take as evidence of learning something about how he learns, as well as about the problem. In the second account, the story is similar except that there was no teacher intervention.

Laurinda: Your learning seems parallel to the learning of the students in these two accounts. There's a recognition of sameness over time, you've made a distinction in your world, then made a shift in the level of communication to talk *about* the awareness.

Alf: Yes, and I'm not sure I would have made the link if I hadn't actually been keeping a diary and writing down the detail of accounts of incidents. I emphasise the importance of writing with my students for exactly that reason, that it's such a powerful way of becoming aware of what you are doing. With my year 7 class at the end of an extended piece of work I get them to write about what they have learnt, both what new mathematics they have learnt and what they have learnt about the process of thinking mathematically.

Evidence box 2.1: What have I learnt?

Working as a mathematician

I've learnt that I've learn lots of things. When I have to think like a mathematician then I try to. I've learnt that you have to think about the promblem and not just do the sum. Also you have to maby carry on thinking about the problem and see if it carries on. You could also have suggestions on why there are problems and why the problem works. It is a lot different to primary school because at primary school we just had to do the sum. We dident think about the problem of the sum we had to just do it.

I've learnt Mathematicions have to think quickly and solve a lot of problems. You've got to jot little theories down. On a lot of theories you have to write why. You've got to correct your mistakes. You've got to confirm things as ABC. You've got to explain your findings.

I have learned that being a mathematisian you have to write your iders down because I used to just work it out in my head. And you have to think a lot harder than useall. And you have to do more work. The questions are harder to make you think more. You get homework not like in a primary school. To be a mathamatison you have to solve a question think how to solve it and why it got solved.

What I have learnt in math since the beginning of term is that mathematicians write things down even though they might not be true and that even the mathematicians see if there statement is true.

We have learnt that you have to work mathamatically to find out problems and "WHY" things work.

What I have learnt

I have learnt that I don't know my tables very well but I think I have improved on them a bit. I have learnt that mathematicians think about why things go in order and why things add up to that number and can I pridict this.

I have learnt about what prime numbers are and a reminder about what factors are. As a mathamatichan I have learnt that to every subject you do you can look for a pattern because then the task you are doing will be a lot easier.

I have learnt about angles and degrese, mathematichions are always thinking how dose it work and why there always ritting down what they do and what they think. I have learnt the triks and how it works you should use N and carry on from there. On the graphs all the lines get steper and steeper.

I have learnt that the first number has to be bigger than the last number in the sum.

We have been writing about the sums and then we know what they mean.

I have learnt how two understand two sums in maths better than I did before

I have learnt its ok to make mistakes

I have learnt how to do 4 + 5 digits better than I did in primary school.

If maths is more exiting in Secondary School than in primarys because in primary school we copyed for texts books and that is SO boring I hated maths but I like it here because you can write on the board and make suggestions and talk about the work and write in our books as I said before

The labels or meta-comments referred to in these pieces of writing include: *think about the problem not just do the sum; find out problems and why things work; look for a pattern; explain things; write things down; solve a question; it's okay to make mistakes; use N.*

2.2 Finding a purpose (Alf)

Reflecting back on my first year of teaching had produced a feeling of inadequacy akin to despair. No lesson really seemed to match up to my ideal image of what seemed possible and there was a strong sense of a gap between where my philosophy lay and the day to day practice of what was actually happening in the classroom.

I remember reading articles in **Mathematics Teaching** [the journal of the UK's Association of Teachers of Mathematics] in my first and second years of teaching and becoming frustrated. They were often about what sounded like wonderful lessons, where the students were being creative and investigative – but I wanted to know how I could even get my students to listen to each other, let alone be creative, and I did not find I got much access as to what the teachers were doing or thinking.

When I met Laurinda towards the end of my first year of teaching, travelling in a car, with my attention partly taken up by driving, she asked whether I could bring to mind particular moments or times during a part or parts of lessons which had felt closest to my ideal.

This provoked two 'brief-but-vivid' (see *Theory Box 2.1*, p. 31) accounts:

Account 3: During an A-Level lesson on partial fractions I was going through an example on the board, trying to prompt suggestions for what I should write. Some discussion ensued amongst the students, which ended in disagreement about what the next line should be. I said I would not write anything until there was a unanimous opinion. This started further talk and a resolution amongst themselves of the disagreement. I then continued with the rule of waiting for agreement before writing the next line on the board.

Account 4: Doing significant figures with a year 9, I wrote up a list of numbers and got the class to round them to the nearest hundred or tenth, [...] Keeping silent, I wrote, next to their answers, how many significant figures they had used in their rounding. Different explanations for what I was doing were quickly formed and a discussion followed about what significant figures were. (Email communication, Alf)

Whilst driving and without any prompting I said, with energy: 'It's silence, isn't it? It's silence.' This process felt like somehow staying with the detail of the accounts and seeing the pattern that was there.

We had found an initial focus for our work together. We jointly planned lessons that would begin with my own silence. I had something in mind (*i.e.* my own deliberate silence) that could inform my planning and actions.

There was a markedly more engaged reaction from some of my classes (see *Section 2.3*). For almost the first time I had a real conviction about my actions in a classroom, linked to my ideals about what I thought was important in teaching.

It seems significant that silence is something I recognised and valued in other contexts. I wrote around the time of the incident above, about my own 'story of silence', an extract of which is below:

> Silence is also in waiting. I think I learned how to wait in Zimbabwe – hours by the side of roads waiting for buses and lifts. I have an image of waiting with Zimbabwean friends who would just sit ...
> (Diary, Alf, 1996)

As Laurinda and I continued to work together during my second year of teaching we identified other foci, or purposes, that I used to inform what I did in my classroom, *e.g.* 'how can I structure this lesson so that I will not be needed by the students?', 'it's not the answer that's wrong', 'sharing responses', 'getting organised'. These labels are fairly meaningless for anyone else, but, for me, focusing on such 'purposes' gave me conviction in my planning and teaching. I was offering this activity or this way of working for a reason that was linked directly to what I wanted to achieve in a classroom.

We would always work together on the mathematics of the problem or activity that we offered. Through doing this, I became more aware of what mathematics I was using or needed. The process of working on the mathematics for ourselves, at our own levels, also seemed crucial in terms of identifying rich questions or starting points. I felt the process, particularly in identifying a range of responses to any starting point, helped sensitise me to what students might say or do and helped me develop some flexibility in terms of my responses. We had a phrase 'planning to conviction' that, for me, was about working to a point of being sure there was enough that was mathematically interesting within my starting point and also having some image of how the students might respond and what I might then do next.

In recognising the power of having some sense of what I was doing that could inform my actions and responses, I began to work on ways of offering students a 'purpose' for a sequence of lessons, that might similarly help inform their decision making. I wrote the following, about one year after working with these ideas:

> A purpose that is articulated to the students is a goal that is not just about the actual activity they are involved with, but places that activity within a wider context. It is a goal that is not necessarily in time and will not necessarily be reached, but it provides a motivation

to engage in an activity, by giving students some sense of where what they are doing fits in to a more complex situation. A relatively global example would be what I keep saying to my current year 11s, that we are working this year at using their mathematics, at trying to solve problems and look at questions that require them to put into practice what they learn. I have my own (parallel) purposes which I may or may not articulate and which, again, are not in time, for example developing my year 11s sense of control over the mathematics they know. (unpublished MEd assignment, Alf, 07/97)

I had observed Laurinda, whilst she was teaching a year 8 (aged 12-13) class of mine, tell the students that their focus for this lesson was: 'not so much on the actual finding of rules but on developing strategies for finding rules'. This offered the students a purpose for that lesson. We planned a series of lessons on drawing graphs of rules during which I issued the challenge (which became a purpose for the students): 'In five lessons time I'm going to come into the room and write up a rule, your challenge is to be able to tell me what the graph would look like, without having to plot it.' Laurinda had used a purpose for students similar to this many times in her own teaching. The idea is one that I found could be adapted to different situations. In these lessons I no longer needed to direct everything that the students did. I was able to encourage students to make their own decisions about what to try out or look at, within the limits imposed by the challenge.

I had written about my first year of teaching that I often felt I offered interactive starts to lessons with no journey. By this I meant that I felt able to use other peoples' or invent my own starts for lessons that could be energising and engaging for students, but somehow I never knew what to do next, except hand out a worksheet of questions for students to try. I would now analyse part of the issue as being that the kind of lesson starts I was using were tightly focused on a specific skill *e.g.* being able to draw and interpret vectors. There was not enough complexity for students to start making decisions or directing their own activity in any way – hence I had to take on the role of providing all the questions.

Sequences of lessons such as the one mentioned above on functions and graphs (see *Section 5.4* for more details) gave me experiences, that were still often isolated, of working for sustained periods with classes in which the students maintained energy for the task and were able to make their own decisions. I believed that many students were learning and coming to realisations about important skills *e.g.* using negative numbers, which they needed to sort out in order to plot the graphs they wanted.

Theory Box 2.1
Brief-but-vivid accounts (Mason, 1996, pp. 25-26)

Armstrong (1980) provides a beautiful collection of accounts of incidents with children aged eight and nine, accumulated as he sat in on the class over an entire year. Few people have time to write even a Pepys-style description of their day, much less take a whole year off to make lengthy observations. Yet it is startling how quickly a vivid moment sinks into the general morass of memory. It can be hard to recontact the moment even a few hours later, unless there is some immediate effort to recall and re-enter it.

What can be most helpful is a brief-but-vivid description which enables re-entry into the moment: to be, as it were, 'back in the situation' at some later time. So all that is needed in a brief-but-vivid account of is enough to trigger recollection. Then the incident can be described to others from that memory, together with any significant further detail that may be needed.

2.3 Using silence – see Brown and Coles (1996)

When we started to work together, Laurinda paid a visit to Alf's classroom in a London comprehensive school. He initially asked the question: 'what do you want to do?' and the only answer from Laurinda was that if the work were to take place the agenda would emerge from conversations. Alf asked the question from a concern that there was not enough of interest in his classroom to warrant such time and attention and was checking out that Laurinda did not have unrealistic expectations of what might be going on. From Laurinda's previous work in teacher development, she was aware that the way teachers talk, describing what they do (*e.g.* 'I use open questioning'), within groups outside their own schools, rarely gives insight into their current practice. It is as if they talk about ideas that they are currently working on which gives no sense of the relative position, for an observer, of what their classroom is really like compared to another class-room in a different school. When a member of one of Laurinda's teacher groups talked explicitly about using open questioning, a visit to the class-room often revealed a quite formal environment which, in the early stages of her work at the RLDU, was surprising. If a teacher was advocating a structured lesson organisation, a visit showed students actively engaged in their learning in an unstructured environment. She had few expectations of what Alf's classroom would be like.

In *Section 2.2* ('Finding a purpose', p. 28) there is a description of how the label of 'silence' emerged. We had found our agenda – we could work on silence. At this stage our concerns were different. Within her theoretical perspective Laurinda recognised 'silence' as Alf identifying a purpose. It would be possible for Alf to work on developing a range of teaching strategies related to this label. Alf was aware that his silence had forced students to think for themselves about what he was doing, putting the onus of explanation on them. He also became aware of 'silence' as a potentially broader categorisation. We discussed teaching strategies involving silence and lesson introductions for algebra, given Alf's scheme of work, drawing on Laurinda's observations of mathematics teachers and lessons. Alf decided what he was going to teach.

Account 5: Silence in year 7

We were going to do some work on arithmogons (see *Fig. 2.1*). I drew one, put a number in two circles, paused and filled in the box in between by adding, paused put a number in the third circle filled in the other two boxes by adding in the same way. A few hands had gone up; there was silence. Another example, still silence; a couple of students bursting to tell the answers in the boxes, but still silence (this was surprising). I was making eye contact with many of the class and looking a lot at a girl who I felt might be the last to pick up what was going on – there was concentration, but still no understanding on her face. A third example ... I turned to look at the class and everyone's eyes were burning into the board – I hadn't experienced this before. Still silence ... I now filled in two boxes with answers from the class, everyone's hand seemed to be up except the girl; she was straining and seemed to have just understood; she half whispered an answer, not quite committing herself – but it was correct. One more example two boys had lost concentration, staring brought them back. The girl's hand was now up with the rest. The boys seemed to be following, so I nodded at the girl and a correct answer came. I then drew an arithmogon with only the boxes filled in and invited the class to try to find what the numbers in the circles could have been, no one needed a further explanation, which is a rare event for me! (Diary, Alf, 09/95)

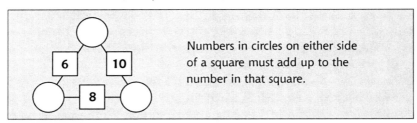

Numbers in circles on either side of a square must add up to the number in that square.

Figure 2.1: The arithmogon waiting for the circles to be filled in

Laurinda was observing the lesson on 18/09/95. In her lesson observation notebook, she recognised the offering and waiting at the start of the lesson as a teaching strategy related, for her, to the label: 'knowing that they all know'. She also reported that there was a 'wonderful build of energy here!' which she recognised as often being an indication that it is time to move on. Speaking into such an energetic space can diffuse the focus that has built up, so the offer of the first arithmogon with the squares filled in, still without speaking, felt smooth in terms of the flow of the lesson.

The story of the lesson illustrates an example of Alf experiencing a change in perceived behaviour of the students in relation to his changed behaviour. The energy of the students in response to his waiting for more of them to put up their hands was what had been surprising.

There is clear evidence of Laurinda also learning. Her recognition of the purpose 'knowing that they all know' refers to Alf demonstrating a teaching strategy of waiting for responses, which had the effect of allowing enough time for more students to be able to offer an answer. At the stage of the 'wonderful build of energy' into the completion of the second arithmogon with circles filled in, however, Laurinda felt uncomfortable. She was aware that she would have made the decision to move on to the next stage of presentation of the challenge of 'what's in the circles given the filled in squares?'. In discussions after the event one story was that this discomfort was felt because Laurinda's experience of teaching told her that either the energy in the students might become dissipated by the task being perceived as too easy, or continue to build and need release in mayhem! Neither of these positions held and all students in this mixed ability group became completely absorbed by the fourth example, after which Alf moved on to the challenging problem.

The teaching strategies used by Alf were:
- giving the students a visual task offered slowly and silently which focused their attention
- timing a change from do-able to a more challenging yet related task.

Having experienced the power of this technique under these circumstances, both Laurinda and Alf were in the position of having their implicit beliefs (Claxton, 1996) brought into question through having experienced a learning episode that they valued for the students. Further, discussion suggested that his belief in 'student autonomy' was an inhibiting factor for Alf in the first year of teaching that could now be adapted as a belief given his increasing ability to take the authority position. For Laurinda it was to see that it was possible to continue the energy build using silence without boring the brighter students for the sake of the 'girl'. In this case the evidence was that the students were not bored and attacked the challenging problem.

Account 6: Silence and energy versus stillness

A year 7 (age 11 – 12) class had been working at seeing whether there was a link between the rules to describe a function and the graphs of the function. Alf had given them a purpose of whether they could know what the graph would look like from the rule without needing to plot the points. The class had been working in the first quadrant only. To introduce these ideas the class had been playing the function game (Banwell et al, 1972, see *Section 1.4*, p. 18). The game is often played in silence.

In discussions before the lesson, the decision had been made for Laurinda to begin by inviting the group to share with her what they had been doing with Alf in the previous lesson. After negative numbers had been introduced into the game, Alf would then refocus the group's attention on the purpose from the previous lesson and extend the work into plotting graphs in all four quadrants. At the handover, Laurinda had used the teaching strategy of holding the silence longer than she would normally have done to force every student to commit themselves to an answer, which in this case was negative. As the pens passed to Alf, the students were excited and present in the task, yet not silent nor easy to handle! It was essential that Alf provided the move from this doable to a more complex yet related challenge so that the energy could be used.

Into an energised space Alf said, 'What's different about what you've been doing with Miss Brown and what we were doing last lesson?' The effect of this question was to make the group absolutely still. This was a silence in the students which was of a different quality to that experienced before – a stillness. (From lesson observation notes)

The silence energy which we had identified in the first visit described above was energetic and mobilising for the learners, their attention was focused in the present and they were using themselves in the moment bringing everything they have with them to restructure their experience. The contemplation (Gattegno, 1987) stillness seemed to be trapped by the invitation to compare two aspects of experience, before and after, and it was as if the students' present attention looked inward to attend to that differ-ence. This stillness was also a powerful silence but not about experience, more about synthesis. There had been the question and this was the reaction – this stillness was their will, not imposed by the teacher.

The interaction of the two researchers/teachers worked creatively as ideas co-emerged and implicit theories of learning and teaching were made conscious allowing possibilities for adaption. There was by now not so much of a feeling of an unbridgeable gap for Alf between his ideals and practice but a sense of staying with the uncertainty and developing ways of working on his teaching.

2.4 Lesson account

Alf's attention in this lesson was partly on working with students on ways of being organised about how they approached a problem in order to give some insight into the underlying structure that might offer the possibility of generalising. Working with the students on their awareness of the need for getting organised in their learning of mathematics was a purpose for Alf, motivating his teaching decisions. There are very few metacomments from Alf in these lesson observation notes but a lot of teaching strategies are evidenced. The lesson is in the second term of the year and so the classroom culture has been established and is developing without the need for many explicit comments from the teacher. When Alf makes a comment that we would interpret as a metacomment or as exemplifying a teaching strategy, often linked to the purpose of 'getting organised', then what he says is italicised.

This year 7 mixed ability class had been set a homework at the end of their last lesson to find how many handshakes would be needed for three people to all shake hands, four people, five people, six, ten, n?

Getting organised – year 7, 1996

– Everyone should have their books open now looking at what they've done ... *as always, one of the difficult problems in mathematics is getting across what you've done.*

Alf invited three students to come to the front and asked: *What is the problem we are working on?*

~ How many handshakes would there be?
~ Shaking hands with each other.
~ Shake each other's hands and see how many times.
~ Everyone has to shake hands with each other once.

Two students directed the three at the front and offered different procedures for ensuring each shakes the other's hands once. Alf invited four different students to come to the front.

– So, *can we get organised here? ... can anyone predict* how many handshakes there will be?

Various answers were called out. Alf invited a student to direct the four at the front. There was some disagreement as to whether the total shakes they got was correct.

– Can we do it so that we can be absolutely 100% sure?

A discussion ensued as to whether 'Dan shakes Simon' and 'Simon shakes Dan' is two shakes or one. A majority of students thought it was one shake.

~ Caroline: With four people you need to do 3 + 2 + 1.

Hayley came to the board to illustrate Caroline's idea. She drew:

Figure 2.2: Caroline's drawing

~ You could do five people using the same system.

~ Just add four to the total.

~ 10, because you add 4 to 6.

– Great, so we've got a prediction for the answer with five people.

Carla came to the board to offer another image for the situation with four people:

Figure 2.3: Carla's drawing

Alf invited five students to the front who took 10 handshakes to shake each other's hands, as predicted. Colin started making predictions for six and seven people by adding 5 and then adding 6. Alf started to draw a table on the board:

No. people	No. shakes	
3	3	) +3
4	6	) +4
5	10	) +5
6	15	) +6
7	21	) +7
8	28	

Figure 2.4: Table of results with first differences

~ Tim: The numbers make triangles.

Tim came to the board to illustrate what he meant and drew:

 *Figure 2.5: Tim's drawing*

~ Colin: Like in snooker.

– This is all very well and this is a very nice pattern but what if there were 15 people? ... *Can we do it more efficiently?* ... *Can we get straight to the answer?*

~ Each person would shake one less than the person before, so B would shake one less than A.

~ The first person shakes 14, then 13, ...

~ Add them up.

~ 14 + 13 + 12 down to 1.

– That's going to take quite a long time especially if we had, say 100 people, *we need a way of working out that sum quickly.*

~ Danya: If you do 15 times 15 then you need to take off one and then take off two ...

Caroline offered the image: 14 + 13 + 12 + 11 + 10 + 9 + 8
                            1 + 2 + 3 + 4 + 5 + 6 + 7

Part of the conversation was as follows:

~ 15 goes to 15 times 14 divided by 2.

~ This is double.

~ Same as 15 times 7.

~ Equals 105.

~ It all equals 15 so 'cos there're 14 numbers you times and then halve.

~ That's the same as what Peter said and then halve it.

~ Caroline: I've got something that I think will help. Can I come to the board? ... You could do it like this: 15 + 15 + 15 + 15 + 15 + 15 + 15, but times is quicker, it's 15 times 7.

– *So how would this work for 100?*

~ You'd be here all day.

– How many does the first person shake?

Murmurs of 99, 98. Alf gave the class *three minutes to try and work out the answer* for 100.

– It looks like there are two or three ways people have tried to work out 100 on your tables, even if you really like your method, *try to hear how others have done it.*

~ Simon: I know a way to do it. ... I'll just show it.

Simon came to the board and wrote 100 → 99 → 98 →

~ Simon: I didn't work it out, it'll take ages ... add them all up don't you?

~ Mr 100 did 99 shakes.

~ It's 50 times 100.

Alf wrote:

$$99 + 98 + 97 + \ldots + 3 + 2 + 1$$
$$\underline{1 + 2 + 3 + \ldots + 97 + 98 + 99}$$
$$100 + 100 + 100 + \ldots + 100 + 100 + 100$$

We agreed this was 99 times 100 but was double the answer we wanted.

~ It's 99 times 50.

~ Is that the same as 100 times 49.5?

~ Colin: Is there any way you could do n in this?

By popular demand n changed to m so that we were thinking about m-people (a pop-group at the time!).

– How many handshakes would the first person have to do? ... If there were 50 people, how many would the first person do? ... *Let's chant this* ... with 50 people how many does the first person shake? ... [49] ... with 250, how many does the first person shake? ... [249] ... twenty people? ... [19] ... m-people? ... [m – 1].

Alf wrote:

$$1 + 2 + 3 + \ldots + (m - 2) + (m - 1)$$
$$\underline{(m - 1) + (m - 2) + (m - 3) + \ldots + 2 + 1}$$
$$m + m + m + \ldots + m + m$$

– How many m's have we got?

~ m minus one of them.

– A quick way of adding up m, m minus one times?

~ m times m minus one.

– But that gives us twice the answer we want, so the total for m people is m times m minus one divided by 2.

Alf wrote: Total for m people = $\dfrac{m(m - 1)}{2}$

– *I wouldn't expect you to do that yourself at this stage but the more you see of it the better.*

(Taken from observation notes, Laurinda, 01/99.)

3 ON PURPOSES

3.1 Realising purposes

Alf: The 1995 interview shows that I was aware of not wanting to be in my classroom simply to confirm correct answers:

> One of the things that has come out in some of the early lessons is when I've been quite explicit about 'I'm not really here to give answers'. That came out in the year 12 class (aged 16-17) at one stage. They were saying 'Is this right? Is this right?' and I kind of resisted. (Appendix 2, 1995, 2.63)

Laurinda: It is common for student teachers, with a strong mathematics background, to want to teach because when they enjoyed doing mathematics it gave them the satisfaction of knowing they had the correct answer, unlike many other essay-related academic subjects. Similarly, student teachers of mathematics can think that their teaching problems will have simple resolutions. Your story of developing your teaching from 1995 to 1999 illustrates the time needed to learn, as any teacher has to, through the process of working with experience, to acquire a range of teaching strategies to adapt to different situations.

Alf: In 1995, I had few strategies to encourage students to explore their different methods for solving problems, but by 1999 I'm aware of focusing students' attention on the issue of learning through helping each other by commenting on it (a metacomment):

> ... and then I said and we're going to be working as a group mathematically and one of the things that that means is that 'there isn't a sense in which there's a right or wrong answer' because it's about learning and you often learn by making mistakes so it's not about getting things right or wrong and if we're going to be working

mathematically as a group it's about helping each other and helping each other understand what's going on. (Appendix 1, 1999, 1.6)

Laurinda: There was still some confusion for you at the start of 1999 around the dichotomy right or wrong and you invited the students to comment on whether they thought something was right or wrong, which is a strategy to get students talking (Appendix 1, 1999, 1.22).

Alf: When I've observed you teaching I am often struck by the fact that you do not seem to view what students do or say in terms of being right or wrong. One instance of this would be *Sarah's Story* in *Section 3.3*. Is this a conscious part of what you do?

Laurinda: I first started observing other people's lessons having left the secondary school where I had been head of department and taught for 12 years. The culture I taught in was one where students were used to exploring within their mathematics lessons. Outside this culture, I found myself distinguishing between lessons by asking: *Is this a classroom in which it's alright to be wrong?* I suppose I wouldn't have been aware of the differences in the behaviours of the students against this question if I hadn't had some sense of students not being afraid of making mistakes in my own classroom. What I think of myself as doing, however, is questioning whether or not I have a picture of what the student is saying and where that has come from and, when I don't, offering something so that we can continue to explore meaning. I have lots of examples of situations where something 'obviously' wrong has proved powerful in uncovering a commonly held misconception, where the student's answer makes sense from their point of view *e.g.,* 'What do you get as an answer to: $1 + 2 \times 3 - 3 + 4$?' A lot of students are taught a rule 'BODMAS' which tells them what to do first in a sum such as this one; Brackets, Of, Division, Multiplication, Addition, Subtraction. One common misapplication of this rule gives the answer 0, since the $3 + 4$ is taken away from $1 + 2 \times 3$. I still ask myself: *Is this a classroom where it's alright to be wrong?* – not consciously now – but I know that it's one of the ways I organise my observations. I am interested in the multiple stories that can be told and the classroom atmosphere seems less rich where the teacher is the judge of students' answers. How do you work with the tension between knowing that sometimes there are 'right answers' and celebrating the differences in interpretation and expression that might be around in your classroom?

Alf: The tension is one I recognise. Working on differences in students' interpretations can create the discussions that seem central to their learning, and yet, as you say, sometimes there are 'right' answers. I have said to a number of my classes this year that I would like them to *stop thinking about answers as being right or wrong*. To say someone has got an answer wrong, there must be a difference in the answer I have and the answer they have. This difference can occur for two reasons; either one of us has made a mistake (and that happens all the time to me) or we have actually been working on slightly different questions *i.e.* we have not been operating under the same assumptions. For instance a student might say that there is no solution to the equation $2x = 3$. You might think this is wrong, since there is a solution 1.5, but the student may have been making an assumption that they had to work with the integers or whole numbers. I think it's actually not so much that I don't believe answers can be wrong, but that my attention is in a different place. Differences in students' answers are an opportunity for learning and I'd like them to come to see what they might think of as getting something 'wrong' in the same way. Looking back at what I have just written, I recognise that *exploring differences in students' responses* is a purpose that I am often conscious of thinking about in my planning. *Writing on the board, without comment, all the different answers to a question and then getting students to talk about how they got their answers*, is one teaching strategy related to this purpose that I know I use. Is that a purpose in your teaching as well?

Laurinda: I do have lots of teaching strategies which I could see as related to 'exploring differences in students' responses', as you wrote about in Section 0.4, but now I think I use a purpose of 'How do I know what they know?' from the start of a lesson and this seems to support teaching strategies where the class works together exploring meaning from some activity which gives them something to talk about. I'm not sure that the purposes during the lesson are explicit at all for me now. I work with what the students bring with them and have complex strategies and mechanisms that are flexible in adapting to what comes. I try not to be surprised or pleased or judge what the students do bring – having high expectations overall but accepting what happens in the moment is another one those tensions.

3.2 On metacommenting (Alf)

A purpose I have worked on in my teaching is to use metacomments both in one to one interaction with students and in wider discussion. In 1996, I defined metacomments as:

> If a student makes a statement which I decide to respond to, I can comment, for example, on whether I agree with the statement or whether I think it is true, or I can offer a similar or different statement

Evidence box 3.1: Making progress by getting organised

I've looked but I can't realy find any thing. All
I've found is that they all got 1 or moor.
I just found that you leve a space then it goes
{1, 2 . . .3} and then leave a space {1, 2, 3} and
it comes on through the factor sheet

Leave 1,(1
(1,2)   2, (1,2)
Leave 3 (1,3
(1,2)   4 (1,2)4
     5 (1,5
     6 (1,2)3,6
     7 (1,7
     8 (1,2)4,8
     9 (1,3,9
   10 (1,2)5,10
   11 (1,11
   12 (1,2)3,4,6,12

se and the pattern goes on
and on through the sheet
and I also notised it goes
1 2 then you get a 3 in the
number. So it would go 2
spaces then on the 3rd
there would be 3 and so on.

               L = Leave

L 1 (1
L 2 (1,2
→ 3 (1(3)
L 4 (1,2,4
L 5 (1,5
→ 6 (1,2(3)6

and so on and so on

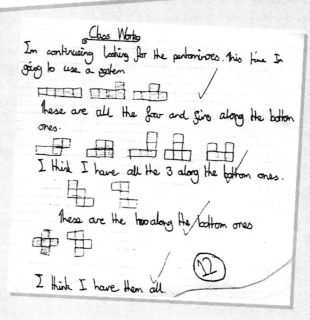

Class Works

Im continueing looking for the pentominoes. This time Im
going to use a system

these are all the four and five along the bottom
ones.

I think I have all the 3 along the bottom ones.

These are the two along the bottom ones

I think I have them all.

12

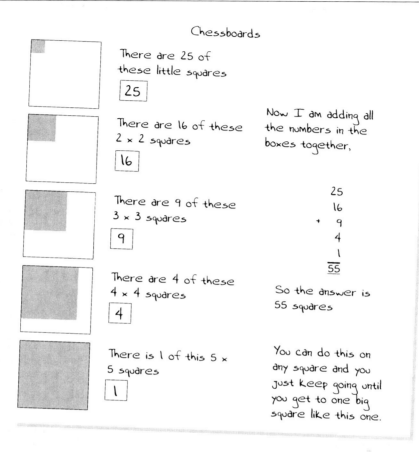

Chessboards

There are 25 of
these little squares

25

There are 16 of these
2 × 2 squares

16

There are 9 of these
3 × 3 squares

9

There are 4 of these
4 × 4 squares

4

There is 1 of this 5 ×
5 squares

1

Now I am adding all
the numbers in the
boxes together,

$$
\begin{array}{r}
25 \\
16 \\
+ \quad 9 \\
4 \\
1 \\
\hline
55
\end{array}
$$

So the answer is
55 squares

You can do this on
any square and you
just keep going until
you get to one big
square like this one.

What I Did

I firts started on the hole ches board but
that was to hard so I done 2 by 2 and
that came to 5 then 3 by 3 and that came
to the answer and I went up to 6 and that
was 54. I think as a estamate I think it
would come to atlest 204
It probably wont but I think it would..

The evidence here is that by 'getting organised' students are able to spot a
pattern or come up with a system for finding all solutions that also allows them
to ask further questions.

about the same subject matter; all of which are responses to the statement itself. I can also decide to comment about the statement, for example 'What is it about this equation that makes you say y is 4.8 when x is zero?' or 'What you have just said is an example of using mathematics to help make sense of a problem, doing that is one of our main aims this year.' These are metacomments, meta because they are at a higher logical level than the content of what was said to me; they are about that content. (unpublished MEd assignment, Alf, 07/97)

We now use the word metacomment in a slightly different way, which I explain at the end of this section. In what follows, however, I am using metacomment in the sense defined above.

I had become aware that my one-to-one interactions with students often seemed to leave them dissatisfied or more frustrated than when they had asked for my attention. Similarly when I ran discussions I often felt they became question and answer sessions with me always asking the question – not allowing students the space to explore the meanings or differences in understandings that they wanted. The pattern of dialogue would consistently go teacher, student, teacher, student, . . .

In deciding to work with metacommenting as a purpose, I initially explicitly limited myself in class to comment only *about* what students said to me and not engage in a content level discussion *e.g.* not get into explaining my understanding of this piece of mathematics but for example to question why they were stuck or whether they had been able to do this question. Examples from one year 12 (aged 16-17) lesson that I tape-recorded are of me saying; 'Has anyone got anything the same or different?', 'Can anyone draw a graph corresponding to Steve's gradient graph?', 'Rita, could you say what Steve meant?'. In each case I am commenting *about* the student's mathematics. In doing so I avoid being the arbiter of what is right or wrong.

Another example of metacommenting occurred when I was observing two students working on a problem as part of a sixth-form conference workshop. The students were working on a problem about creating a curve of a particular length and were talking to each other about their ideas but seemingly not making any progress. It became clear to me, listening to them, that they were using the word 'length' in different ways. I commented 'I think you should talk to each other about what each of you means by length in this example'. This was enough to provoke them into a discussion that allowed them to continue working together on the task.

I was running a session at the same sixth-form conference a year later in which a group of twenty students were discussing the solution to a problem involving two runners who dead heat in a 100 metre race. The question was: 'Must there be a point during the race at which they are travelling at the same speed at the same time?'. There was a discussion about a solution that a student had presented to the group. Someone commented that the runners had not necessarily run the same distance, to which the girl presenting the solution replied that they must run the same distance. I recognised in that moment an opportunity for a metacomment and I suggested *both students talk to each other about what they meant by* 'distance' which provoked them to recognise and agree with what each other meant.

This second example was striking to me because, in recognising the situation as being almost identical in form to the incident the year before, I realised that metacommenting had gone away as something I was working on explicitly in my teaching. In reflecting afterwards on the discussion up to that point I actually had been metacommenting at times *e.g. asking one student to say what they interpreted another one as having said* and *asking a student who didn't understand what another had said to ask them a question*. Working on metacommenting as a purpose had meant such behaviours were now automatic. I was unconsciously using metacomments in running this discussion but not as consistently as when I was doing this consciously. Metacommenting subsequently returned, for a time, to be something I thought about more explicitly in my teaching.

I would now interpret the examples above in a slightly different way. I actually seem to be talking about two different levels of comment. The first, *e.g.* 'What is it about this equation that makes you say y is 4.8 when x is zero?', is an initial shift in level that serves the purpose of reflecting the student's comment back to others in the class and is a comment about the student's mathematics or an invitation for other students to act. I would now call this a teaching strategy. The second level of comment, *e.g.* 'What you have just said is an example of using mathematics to help make sense of a problem, doing that is one of our main aims this year', represents a further shift and is a comment *about* a comment that a student has made *about* the mathematics. I almost always relate these second level comments on student behaviours or comments to the task for the year (purpose) of becoming a mathematician. It is these second level shifts that we call metacomments in this book.

3.3 Sarah's story – see Brown and Coles (1997)

What follows, after a brief background to the incident, is an account of what we call Sarah's story. We then give two interpretations of this story to give a detailed account of how purposes, teaching strategies and metacomments are an integral part of the way we work as teachers and researchers.

Background

The mixed ability class of 11- and 12-year-olds were in the middle of an investigation related to perimeter and area. They had all started with the problem of finding the rectangle with the largest area, given a perimeter of 12cm. Having solved this starter problem students were encouraged to try other perimeters, begin looking for patterns and start generating and working on their own questions. At the beginning of one lesson we shared what they had found out so far:

- The largest area for any different perimeter is a square.
- To draw a rectangle with a perimeter of an odd number you must use halves.
- Odd sides means an odd area.
- Even sides implies an even area.
- Divide the perimeter by 4, then times by itself, what you get is biggest area.

These statements were written on the board as they were said and no explanations were asked for or given. When the list was completed, Laurinda added that she had been thinking about the first question and wondered whether it was possible to find the perimeter of a square when the area was 50cm². There was an invitation to stay with what they were working on or incorporate any of these ideas into their investigations. The class continued to work individually or in small groups.

Account 7: Sarah's story

After some time Sarah, who had been working on the 50cm² problem, came up to Laurinda stating that the perimeter must be 4cm. Laurinda drew a square with area 1cm² on the board with 1s marked around the perimeter (see *Fig. 3.1*) and waited for some response.

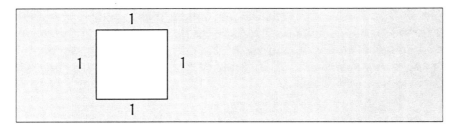

Figure 3.1: The sketch of the square shown to Sarah

When none came Laurinda asked Sarah how she had worked the 4cm out. Sarah talked about 'reversing the rule *divide the perimeter by 4 then times by itself to get the biggest area*'. Laurinda started writing the flow diagram for this as she spoke and Sarah reversed 'times by itself' to 'divide into itself' and 'divide by 4' to 'times by 4'. This gave $50 / 50 \times 4 = 4$.

Laurinda said, 'OK, times by itself' and wrote: 4 →

Sarah quickly replied '16' having had experience of writing functions in this way, and Laurinda wrote 16 next to the arrow, followed by: 6 →. The answer '36' again came quickly. Laurinda offered: ← 49

Sarah responded with 7 immediately. They agreed that this had not been 'divide into itself' but what was it? Sarah went off to work on this.

First comment: It's not the answer that is wrong!

Why did Sarah come up to Laurinda with her offer of 4cm for the perimeter of the square of area 50cm²? The question was posed by her and perhaps Sarah was wanting to show that she had a solution? Or perhaps Sarah was aware that the perimeter could not be 4cm and she was wanting to sort this out? We cannot know. We are not concerned with Sarah's possible motivations. As researchers what we keep coming back to is the detail of our actions as teachers and the students' responses.

Reflecting on what happened, Laurinda did not dismiss Sarah's offer of 4cm as being wrong, nor tell her what the answer should be, nor immediately try to understand her thought processes. Instead she offered a square with perimeter 4cm (*Fig. 3.1*) which was not responded to. Laurinda also did not try to find out why Sarah did not respond. It would be possible to make interpretations about why Sarah said nothing. Perhaps she was already aware of this conflict? Perhaps she could not connect the image with her way of getting to the answer of 4cm? Laurinda let this go since even a directed question to try to get Sarah to work out the area of the square with perimeter 4cm seems to be taking Sarah further away from her own thoughts.

Why not ask straight away how she had got 4cm? Laurinda's initial offer had come from sharing her own understandings of what a square of perimeter 4cm would look like. This was creating a conflict for Laurinda: How could a square of perimeter 4cm have an area so much greater than one? Her prime interest is not in how Sarah got 4cm at this stage. In working with students interactively, if she recognises that she does not accept what has been said she shares with them what seems to be creating her problem. Given that the offer of the square did not take Sarah any further, Laurinda needed some more information so that another offer might be possible. Laurinda now asked how Sarah had got 4cm. This provoked an energised statement from her 'reversing the rule *divide the perimeter by 4 then times by itself to get the biggest area'* which seemed to be saying 'This has to be true – even though it feels odd'.

Here the student is convinced of their world. Laurinda has something to offer which came from her awareness that times by itself and divide into itself are not inverse operations. The offer came in a form that was not telling. Laurinda was offering objects in the world for Sarah to adapt to. Sarah had a strong awareness of inverse (doing and undoing) but, as it turned out, needed to extend those ideas to cope with squaring and square rooting positive integers. At a later stage she might meet a context where 49 might demand a different response *i.e.* one that included →7.

This incident revealed that Laurinda's implicit belief in working with students is that *It's not the answer that is wrong!*

We have since explicitly developed a practice of never commenting on answers as being wrong since they have arisen from the student making sense of their world. We work with the process, offering more complexity for the student to adapt to in some way and offer our own images when these are in conflict in some way with theirs.

What is motivating Sarah to come and interact with Laurinda? How does Laurinda know what to respond? Laurinda wrote at that time:

> No two events or responses are ever quite the same in the classroom. In contrast, when I begin to work on a new piece for the piano I first attend to the fingering in detail and practise difficult transitions. Each time I play the piece the fingering will be the same; eventually the fingering is automatic. There is little in the detail of our practice of the teaching and learning of mathematics which is exactly repeatable in this way; no one has come up to me with Sarah's question before.

What does seem repeatable is on the level of purposes. For instance, at the start of a topic or theme how can we find out what the individual students in our class know and where they find problems so that we can make decisions about what to offer? We, as teachers, would have a whole variety of possible strategies we could choose to adopt to carry through such a purpose. Which one we use would depend on the individual circumstances of the class.

Second comment: 'It's not the answer that is wrong!' as a purpose

> In Laurinda's actions during Sarah's story we recognise her purpose of not commenting on answers as being wrong. We are not simply telling what is right from our own viewpoint but are moving away

from the right/wrong dichotomy into something richer and more complex. Purposes help us to deal with the decision-making necessary in the face of the complexity of the classroom. Laurinda did not know that these particular circumstances would arise but became aware that a purpose she always operates with is *it's not the answer that is wrong!* Laurinda was aware in the moment of her behaviours in the face of Sarah's statements, she heard what Sarah was saying and responded to that. The purpose 'it's not the answer that is wrong!' is the distillation of a complex web of intentions, thoughts, past experiences and actions that inform practice. (Brown and Coles, 1997)

In preparation for the lesson in which the Sarah incident occurred we had been explicitly focusing on the purpose of *sharing responses*. Alf, as a teacher, was interested in extending his repertoire of strategies to explore the richness of responses among the students. Laurinda was working with how a model for teacher development, related to purposes, functioned in allowing Alf to work on this. As the students in the story worked on their own questions or ones that had been written on the board, they too were making decisions about where to take their investigations. They were exploring an increasingly complex field of ideas about area and perimeter.

Sarah was working with energy on this mathematics. She was 'mathematising' (see *Theory Box 3.2*). Whilst sorting out relationships between area and perimeter she was identifying particular aspects to focus on such as 'what's the perimeter of a square with area $50cm^2$?' and 'what's the opposite of times by itself?'. We believe that part of what was motivating Sarah was the fact that these were her questions that she had chosen to engage with within the wider problem set to the class by the teacher.

What seems important for all the purposes for teachers and students is the way they motivate learning. Sarah moved between her questions and the actions of accessing mathematical techniques that were needed. The teacher moved between the need not to see answers as simply wrong and the actions or teaching strategies for achieving this that had been developed over time. Purposes are very clearly linked to a range of possible actions. They are removed in some way from the current action but provide an organising strand, often over a long time, for learning through experience and also support the decision making necessary for the individual to act.

The research process of interacting theory and data with the re-tellings of the stories from our practice over time is an essential part of our work since in itself the creation of narrative drives us forward and allows us to act.

Theory Box 3.2
From Wheeler, D., 1975, Humanising mathematical education.
In *Mathematics Teaching* No.71, p4-9

The substitution of the goal of facilitating children's mathematical activity for the goal of passing on mathematical knowledge is a substantial step in demonstrating that mathematics is a human activity. Activity is personal whereas knowledge often seems impersonal; activity is dynamic whereas knowledge is frequently inert; activity implies involvement in one's own learning rather than passive acceptance of someone else's.

But this is not any longer a new message and perhaps we need to consider how to avoid the danger that mathematical activity becomes a label for something too diffused and generalised, a way of learning in which almost anything goes. It may be another step, if only a relatively small one, to substitute for the encouragement of mathematical activity an education which zeroes in on mathematisation (an ugly word, but no matter). The shift of emphases can take us even further away from an exclusive reliance on external criteria of quality derived from the mathematics of the past. Even though the aim to promote mathematical activity was designed to stress the importance of the 'process' over the 'product', we have tended to reassure ourselves that what we were encouraging was actually mathematical activity by making sure that the product was recognisably familiar mathematics. So, in a way, the nature of the product still dominates our judgements. On the other hand, the word 'mathematisation' is a label for the process itself and provided we can back up our use of it by a strong rational conviction that children have the necessary functionings to be able to mathematise, we may be liberated from the tyranny of judging mathematics only by looking at what it has produced in the past. . . .

In a crude attempt to make explicit the nature of mathematisation, I would include the following ingredients: the ability to perceive relationships, to idealise them into purely mental material, and to operate on them mentally to produce new relationships. It is the capacity to internalise, or to virtualise, actions or perceptions so as to ask oneself the question, 'What would happen if . . . ?'; the ability to make transformations - from actions to perceptions, from perceptions to images, from images to concepts, as well as within each category – to alter frames of reference, to refocus on neglected attributes of a situation, to recast problems; the capacity to coordinate and contrast the real and the ideal and to synthesise the systems of perception, imagery, language and symbolism. When these functionings are applied to pure relationships, detached from specific exemplars, the products will then be mathematics.

For a presentation of one way to generate games and activities for children that engage their powers of mathematisation in order to get them to produce elementary mathematics of a familiar sort, I refer readers to Caleb Gattegno's (1974) *The Common Sense of Teaching Mathematics*. (pp.5-6)

3.4 Lesson account

What follows is based on notes taken by Laurinda at a lecture Alf gave to sixth formers (aged 16-17) in South Gloucestershire on the 4th July, 2001. The title of the lecture was: *Some things mathematicians do for a living ...* Laurinda, as an observer and recorder of the session, was struck by how Alf worked with these students in a lecture theatre in a similar way to how he works with year 7 (aged 11-12) students in his classroom (see Appendix 1). The similarity lay in the commenting by Alf on the students' ways of working and thinking mathematically rather than on the mathematics itself *i.e.* in his metacommenting. Ways of working and thinking, in an environment where Alf knew none of the students previously, were established from the start of the lecture. The metacomments provided the students with clear messages about how the lecturer wanted them to respond. There is evidence that Alf was doing this in awareness given that the title of the lecture made no reference to its mathematical content.

Metacommenting – year 12 (aged 16-17), 2001

The students arrived and the first overhead, which was the title of the lecture, was displayed. Alf made a comment about how he hoped that why he had chosen the title would become clearer as they worked on some mathematics together. The dialogue is typed from Laurinda's observation notes. She took down as much of the detail of what then happened as possible. What follows is a series of snapshots from the lecture. Alf's metacomments, that Laurinda was aware of at the time, are indicated by italics. When Alf makes a comment that we would interpret as a teaching strategy, often linked to a purpose of *opening up and slowing down discussion*, then what he says is also italicised. ... indicates a pause and [...] some missing dialogue or text.

> – Visualise a cube ... each corner, each side, each face ... count the number of corners, or vertices ... the number of edges ... and the number of faces; V, E, F. Work out what $V - E + F$ equals ...
> – *Can you talk us through your answer?*
> ~ 8 corners.
> – Can you say how you got that?
> ~ 4 round the top and 4 round the bottom.
> – Now edges?
> ~ 12. 4 on the top face, 4 on the side and 4 on the bottom.
> – *Can anyone say that in another way?*
> ~ 4 on the top, 4 on the bottom and 4 joining.
> – Faces?

~ One on the top, one on the bottom and 4 round the side.

– *Ask a question* … anyone not happy with that?

Alf then displayed the table (see *Fig. 3.2*) on a whiteboard and wrote in the answers for the cube.

– So, what can we do next?

~ Go through different shapes and look at patterns.

Shape	V	E	F	V−E+F
Cube	8	12	6	2

Figure 3.2: Table drawn early in the lecture

– *One thing mathematicians do for a living is search for patterns* … What other shapes could we try?

~ Sphere. (Laughter.)

– Why the laughter?

~ Dodecahedron.

– *Can you describe?*

~ Lots of faces, vertices and edges.

~ Like a football but not so spherical …

~ Made out of hexagons and pentagons. Pentagon top and bottom and hexagons around the sides.

– This seems quite hard for us all to think about. Could someone suggest another shape *we could work on visualising or drawing?*

~ Tetrahedron.

– *Everyone visualise* … what's V, what's E, what's F?

~ Vertices, 4

~ No, vertices, 5, 4 plus 1

~ 5 is a square-based pyramid.

~ But a tetrahedron is a triangle-based pyramid.

– Where are the four?

~ One at the top, three around the bottom.

– Edges.

~ Three around the bottom of the pyramid and three between the side faces.

– How many faces?

~ 4

Alf also wrote these results on the table.
    – You talked about patterns. Have you any?
    – *One thing that mathematics is about is asking questions.*
    – Anything else? What have you noticed?
    ~ You get the same for decent shapes.
    – *One thing that mathematicians do for a living is classify.* Some it works for, some it doesn't. More things?
    ~ They all go into each other.
    ~ They all have the same common factor.
    – That's what common factor means.
    ~ Why is the answer always two?
    ~ Would you get a different answer doing different shapes?
    – *'Why?' is a question mathematicians ask all the time.*
A student suggested looking at a triangular prism and, after some discussion, Alf wrote the values of V, F and E on the table.
    – Extend the prisms, hexagonal, pentagonal ... Try some more shapes yourselves ... checking out whether $V - E + F = 2$
    – The people who are at the front or at the sides could *come and write up what they find on the board*. Put your hand up if you're in the middle of a row and I have two helpers who will come and write on the board for you.
    By the end of this working session the table is filled in down to octahedron (see *Fig. 3.3*). Four students worked in a group putting two square-based pyramids and two triangular-based pyramids together. They did not know the names of the solids.

Shape	V	E	F	V−E+F
Cube	8	12	6	2
Sphere			1	1
Dodecahedron				
Tetrahedron	4	6	4	2
Sq-based Pyramid	5	8	5	2
Triangular Prism	6	9	5	2
Hexagonal Prism	12	18	8	2
Cylinder	0	2	3	1
Pentagonal Prism	10	15	7	2
Hexahedron	5	9	6	2
Octahedron	6	12	8	2
n-gon Prism	2n	3n	n+2	2

Figure 3.3: The filled-in table

– Any comments?

~ For the prisms the numbers of vertices was twice the number of the shape on the side.

– *Mathematicians generalise.*

– I'm not sure what to call this. (Alf wrote down n-gon prism.) V is 2n.

– Faces?

~ Just for the prisms the edges are 3n.

– Could *anyone explain why* there would be 15 edges for a pentagonal prism?

Alf drew what was said on the board: 2 pentagonal faces, 5 other faces and 5 edges joining corresponding vertices.)

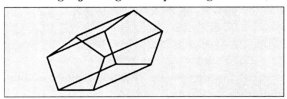

Figure 3.4: Alf's drawing of a pentagonal prism

– Faces?

~ n + 2

- What's V − E + F?

~ 2

– *Something mathematicians do for a living [...] conjecture [...] proof. Has that argument convinced you? Am I convinced by someone else's conjecture?*

– Anything else?

~ Pyramids get half numbers and round it up.

– Does it work for the tetrahedron and the square-based pyramid?

Some discussion was lost here. Alf moved the discussion on without a resolution of the general form for the pyramids.

– *Would you get a different answer for different shapes?*

~ Spherical shapes have no points.

~ From what we have so far it goes to one.

~ Conjecture: For shapes with vertices, V − E + F = 2

– *I have to make a decision. How much time do I have?* I have twenty minutes. We could do some work on the *why* question, why is it two. This is called Euler's rule. Go and look that up on the web. I'll recommend a book at the end of the talk (Lakatos, 1976).

– What we are going to do is look at things the other way around. Assuming the formula works, what does this tell us about what other shapes are possible.

– Mathematicians look at simpler cases. You know that already from your coursework. So, let's look at triangular faces. Three triangles meeting at every point is the simplest shape.

Alf displayed the OHT (*Fig. 3.5*):

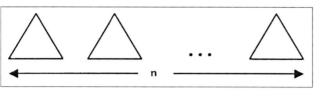

Figure 3.5: n triangles

– I've got these n triangles, how many vertices have they got in total?

~ 3n

– How many edges?

~ 3n

– How many faces?

~ 3n

When these triangles are put together, vertices are lost. There was a discussion of losing vertices through subtracting 2 and dividing by 3.

– V = (3n)/3

There was discussion about edges sharing 2 triangles, hence dividing by 2.

– E = (3n)/2

– ... and F = n

Alf wrote on the board:

> V − E + F = (3n)/3 − (3n)/2 + n = 2
> So, n − (3n)/2 + n = 2

– What's n?

~ 1 and a half

~ 4

– Can you tell us how you got that?

~ n + n = 2n; (3n) divided by 2 is 1·5n

Alf wrote: 2n − 1·5n = 2

~ 2 − 1·5 = ·5

Alf wrote on the board: 0·5n = 2

~ n = 4

Four triangles were given to a person in the front row and they built a shape with three triangles meeting at every point.

– We started with three triangles at a point.

– We are being systematic about this.

– Now try 4 triangles meeting at every point.

There was discussion that in this case, $V = 3n/4$, $E = 3n/2$, $F = n$

– What about 5 triangles meeting at a point, 6 triangles? Try squares, 4, 5, 6, squares meeting at a point. What about pentagons or hexagons? See if you can reproduce the same argument as we've gone through on the board, what will n be in each case?

The students were then given time to work in groups. When a group or individual found a result it was written on the board. At the end of this period, the whole group had found that, with 3 triangles meeting at a point, n=4, with 4 triangles meeting at a point, n=8, with 5 triangles meeting at a point, n=20, and with 6 triangles meeting at a point, n=0. There was some discussion of what this result meant. Alf closed the lecture by inviting the students, in their own time, to try and prove that there could only be 5 Platonic solids, *i.e.* solids whose faces are all regular polygons. He also mentioned that in 4 dimensions there are 6 Platonic hyper-solids and in all higher dimensions there are only three. Again, this is something students could explore via the web if they were interested.

4 ALGEBRA

4.1 About algebra

Alf: In 1995 I was used to working with my students on generalising in mathematics but I think there was a sense of one method of solution where the task for everyone was to 'get' that solution. When students had difficulties I would help them with that one method:

> ... there was one girl I spent a bit of time with ... By the end she'd just about, I'm not sure she did get it absolutely right, but she'd kind of got it. (Appendix 2, 1995, 2.46)

Laurinda: By contrast, by 1999, it is clear that you were thinking about teaching mathematics to students so that they could have an awareness of the structures which underpin it. For example I've seen you use activities supported by the following grid of numbers (Gattegno, 1989, p.29):

0.1	0.2	0.3	0.4	0.5	0.6	0.7	0.8	0.9
1	2	3	4	5	6	7	8	9
10	20	30	40	50	60	70	80	90
100	200	300	400	500	600	700	800	900

Alf: I often use this grid to work on multiplication and division by powers of 10. If it was multiplication by 10, I would point to 6 and the class chant back 60. I point to 30 and then 7 and the class chant 370 (see Appendix 1, 1999, 1.36). Another activity starts with chanting the complements to 10 of numbers, for example, I say 6, students chant back 4, I say 8 and students chant back 2. Having spent time until the class are confident with complements in 10 I will always move to complements in 100 or 1000 or 1.

Laurinda: Like you describe in the transcript from 1999:

I said you're all so good at the 10s we'll try 100. And they were all 'Oh, no' but then it was easy (60 – 40) and I said, so why is it so easy? They said it's just the same, you add a nought on the end and someone said, 'you can do 1000.' (Appendix 1, 1999, 1.42)

Alf: I now see activities such as chanting complements to 10, 100, 1000 as algebraic. I think for me algebra underpins the whole of what I would call mathematics.

Laurinda: When I was at secondary school we were taught algebra, geometry and arithmetic separately. We even had separate exercise books, purple for algebra, green for geometry and orange for arithmetic. When I started to teach ideas had moved on and I saw myself as teaching mathematics. I'm not sure I would have seen algebra as underpinning everything I did. I was aware that the three strands existed in mathematical activity and worked, when teaching, to make sure that in any activity there was something on offer in each strand. The separate subjects seem to be back in fashion with the current National Curriculum. I suppose the techniques of algebra were shown to turn most students off the subject of mathematics and that led to there being less and less algebra content in the syllabuses. On a committee I was on looking at the teaching and learning of algebra in secondary schools, a key conclusion of the report (Sutherland, 1997) was that: 'the National Curriculum is currently too unspecific and lacks substance in relation to algebra'. In the new framework for mathematics at Key Stage (KS) 3 algebra is given a higher priority again.

Alf: When I started teaching I think algebra was often an endpoint of activities, for example, a rule that some students would get to. In a problem solving situation, once we had an algebraic description, it seemed to me there was nothing more to do. I was struck by how this was not the case in the lesson that I saw you teach my year 8 in London (see *Section 1.4*)

Laurinda: I was working with the students on the equivalence of their different algebraic descriptions of the same situation.

Alf: Around that time we came across a description of algebra as 'holding the process' that I found useful: $x + y$ is holding the process of the addition of whatever x and y are. The statement, $x + y$, is, as it were, delaying the resolution into a numerical answer (Gattegno, 1989, p.40). I found in my teaching that students seemed much more able to deal with and use algebraic statements when they had spent time working numerically on the process that the algebra 'held'. For example, I had positive experiences of problems such as 'I think of a number, multiply by 3, add 4 and get 19 – what number did I start with?' leading to students being able to read and manipulate complex expressions (*e.g.* $2(5x-6)=52$ would be translated into: 'I think of a number, multiply by 5, take 6, multiply by 2 and get the answer 52'). What seemed to be important was that the algebraic symbols arose out of a meaningful context. I was still seeing algebra as some sort of endpoint however.

Laurinda: Is that how you see algebra now?

Alf: My attention shifted from wanting students to be able to manipulate algebraic expressions to wanting them also to appreciate the power of symbolism. I began to view algebra more as a way of thinking, a stepping back from the process I am involved in, to be able to express, communicate and so make choices about that process. When I was awarded a teacher-researcher grant from the Teacher Training Agency (TTA) it was to explore working with a year 7 class so that these ideas were around right from the beginning of their time in secondary school.

Laurinda: I remember the challenge we set ourselves for the TTA year: 'Can we develop a school algebra culture in which pupils find a need for algebraic symbolism to express and explore their mathematical ideas? (Sutherland, 1997 p.46). The vision was of a classroom where students are used to doing mathematics that means something to them and where they attempt to express what they are aware of. Their need for algebra would come from their expression of awarenesses within complex situations. At that time I had been exploring my own challenge of students seeing algebra do something for them that they couldn't do without it. This meant, for instance, seeing a proof of something that they knew happened in many cases. The algebraic power of seeing all cases being dealt with felt separate to the collecting of numerical information.

Alf: Before applying for the TTA grant, I had done a literature search on past research about algebra. I needed some way to help me think in more detail about what algebra was. To observe and report on whether my students found a 'need for algebra' we needed some way of noticing and classifying what they did, like your distinction between proof and collecting numerical information. I think an important stage in the research process was finding a definition that talked about algebraic activity as opposed to algebra.

Laurinda: What's your current interpretation of that definition?

Alf: The original definition differentiated between three components of algebraic activity; generational, transformational and global meta-level (Kieran, quoted in Sutherland, 1997, p. 12). Collecting numerical information is a 'generational activity', where symbolism comes from generalising a pattern, say, or expressing a numerical relationship with a rule. If I do any simplifying or manipulating of algebraic expressions, that would be 'transformational' activity (this is about the only aspect of algebra that I was taught in school). The power of mathematical proof, dealing with all cases at once, feels like a 'global meta-level' activity where you are working on problem-solving, explaining, justifying and need an awareness of the structure of the situation you are dealing with. Over the course of the TTA year I found it useful to distinguish between these components in what students did. I was most interested in promoting global meta-level activity since this is the component that seemed tied to 'need'. Problem-solving, justifying, awareness of structure feel to me at the heart of all mathematical activity. It is in this sense that I see algebra (which does not necessarily entail the use of letters) as underpinning the whole of mathematics. Chanting complements to 10, 100, 1000 forces an awareness of structure that feels algebraic in this 'global' sense.

Laurinda: Within the discussions of the committee I was on, Kieran's (1996) definition was the one which included every member's own interpretation of what algebra was. Such a broad definition did seem to allow us to work on our recognition of what algebra was in what the students did with fresh eyes and we started to talk more about algebraic activity rather than using the word algebra.

Alf: I think algebraic activity also interested us because it has the sense of shifting level of abstraction which has to keep grounded in meaning whilst freeing us from it. And then we started to think of algebraic activity as 'natural' since it was always present and within contexts laden with meaning students were manipulating expressions without any formal instruction in transformational techniques.

Laurinda: How would you describe your role as a teacher in all this?

Alf: As a teacher, I now believe that algebraic activity underpins the whole of mathematics. Awareness of the mathematical structure behind what we are doing helps guide my decision-making. I want my students to be looking for structure as well. So, part of my role, when the students are not thinking structurally is to judge when it is appropriate to offer something to force their awareness. An example of such an offer is *Fig. 4.5 (Section 4.4)*, an image which for me demonstrated an underlying infinity.

Laurinda: Structure makes me think of algebraic structures and ideas such as inverse and distributivity, say. These are part of Kieran's (1996) list of global meta-level algebraic activities alongside ways of working such as problem-solving and proving. All these aspects are now part of the culture of your classroom and support the mathematical atmosphere. With this broad definition, algebraic activity is part of everything your students do.

4.2 Becoming a mathematician (Alf)

Sarah's story (*Section 3.3*) is a description of an interaction between Laurinda and one student. I was in the classroom at the time this occurred. It was an incident Laurinda and I came back to talk about many times. My interest in what happened came from an awareness of difference – Sarah had left that interaction in an energised manner and knew what she was going to do next. I was aware that my one to one interactions with students often left them confused about what to do next and with less motivation to sort out any problem that they had before we had started talking. I wrote at the time:

> I can see myself now blundering into a 1-1 interaction with a student and taking out any energy they had about the problem. (Diary, Alf, 01/97)

'Blundering' into an interaction carries images for me of having my own agenda about what I wanted the student to do, and not actually listening to what they said.

The diary entry above was written as part of an exercise I did at the start of a Masters in Mathematics Education course at the University of Bristol, Graduate School of Education. I began the course in my third year of teaching having also just moved to teach in a Bristol school. I identified my one to one interactions with students as the first research issue I wanted to work on. I began tape-recording lessons and looking at transcripts of incidents with students. Times when I was clearly not listening to what had been said to me were painful to hear.

After a few weeks I made a connection that the quality of my interactions with students seemed linked to the wider context of what was happening in the lesson – *i.e.* to my planning. I wrote in my first Masters assignment about one lesson I had tape-recorded:

> What happened in the lesson left the majority of students unable to participate and my own lack of clarity as to what I wanted left me unable to hear, or adapt to, what was said. Was I really prepared to stay with the questions the students were working on or, despite having set up an open activity, do I actually have a more fixed idea of what I want them to do? The evidence suggests the latter, hence there was a conflict between what I had ostensibly set up and

Evidence box 4.1: Students' developing use of symbols

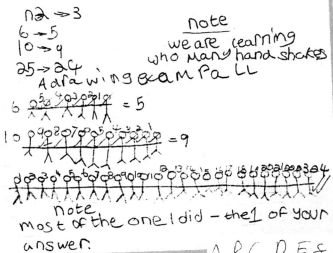

n2 → 3
6 → 5
10 → 9
25 → 24

note
we are learning
who many hand shakes

A drawing exampall

6 = 5

10 = 9

note
most of the one I did – the 1 of your answer.

A B C D E F G H I

These two pieces of work relate to the 'handshakes problem', ie, how many handshakes are needed for everyone in a room to shake everyone else's hand?

Octagon   Septagon   Hexagon
7×5 + 1 = 36   6×4+1= 25   5×7+1= 36
(Y)   (Y)   (Y)
8+7×8 ⊕ 7y+1   6y+1   5y + 1
Zetersqharagon
4 X Y+1= 25
(6)
4Y+1   (QY+1)

Q = One less than amount of matchsticks
Y= amount of shapes

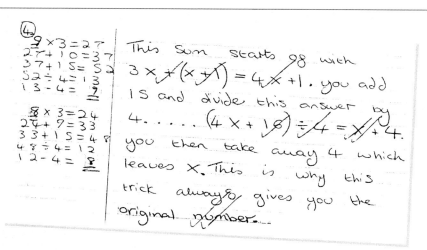

(4)
9 × 3 = 27
27 + 10 = 37
37 + 15 = 52
52 ÷ 4 = 13
13 - 4 = 9

8 × 3 = 24
24 + 9 = 33
33 + 15 = 48
48 ÷ 4 = 12
12 - 4 = 8

This sum starts with $3x + (x + 1) = 4x + 1$. you add 15 and divide this answer by 4 ..... $(4x + 18) ÷ 4 = x + 4$. you then take away 4 which leaves x. This is why this trick always gives you the original number.

HW   10th October

[□ □ □] → 100   this is the rule for any number not just 100

a very nice way of visualising the problem $3 × (x) + 1$ = answer
x = Number of boxes in the line.

N+3 means A Number add 3
0.5N means a half multiply by N

N ÷ 2 means a half of N
N ÷ 3 means a third of N
N ÷ 4 means a quarter of N
√N means the square root of N
N/2 means half of N
N^2 means N multiplied by itself, N × N
N^3 means N × N × N

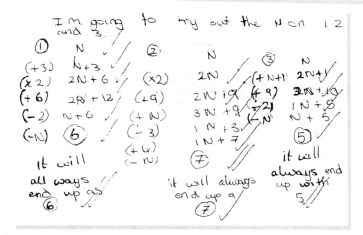

I'm going to try out the N on 12 and 3.

(1)   N
(+3)   N + 3
(×2)   2N + 6
(+6)   2N + 12
(-2)   N + 6
(-N)   6

It will all ways end up as 6

(2)   N
(×2)   2N
(+9)   2N + 9
(+N)   3N + 9
(÷3)   N + 3
(+W)
(- N)

(7)

it will always end up a 7

(3)   N
(+N+1)   2N + 1
(+9)   2N + 10
(÷2)   1N + 5
(-N)   N + 5

3N + 9
1N + 3
1N + 7

(5)

it will always end up with 5

Evidence from across the attainment range of students shows that they are devising their own symbolism to help them think about problems. As with "Getting organised' this is clearly allowing them to make progress in solving these problems.

Evidence box 4.2: Students' work: Natural algebraic activity

Nathan's conjecture: If T + S equals L + R then there is an unlimited amount of solutions.

Liam's conjecture: If T + S does *not* equal L + R then there are NO solutions.

Algebra made my conjecture turn into a theory.

$N \rightarrow NN$

$N \rightarrow 2N + 2$

$N \rightarrow N + N + 2$

$N \rightarrow 2 \times N + 2$

$N \rightarrow N \times N + 2$

$N \rightarrow 2 \times (N + 1)$

$N \rightarrow 2(N + 1)$

I have learnt that being mathematical you have to ask questions from finding out one thing leading to another. You have to ask why all the time or how because you want to find out more. Chris found a conjecture and I asked how? and why? I noticed a difference, and then noticed the difference of difference.

Student:	How could you draw it?
Teacher A:	Well, it would be a sixth of a unit. Very small.
Students:	If you drew it really big so one square was 6 ...
	Sir, what would just a straight line be?

A collage of students' comments and written work to illustrate the mathematical interaction. In being mathematical the students are inevitably engaging in algebraic activity at a global-meta level (see *Section 4.1*) as they justify and prove in response to the question 'Why does it work?'. Transformational and generational algebraic activity is then part of a seemingly natural process.

---

communicated as the aim of the lesson and what my actual agenda was. (unpublished MEd assignment, Alf, 07/97)

I had been working with the notion of giving the students a 'purpose' for a sequence of lessons and encouraging them to ask their own questions. It seems from this writing that what I communicated to students was at times in conflict with my own purposes for the lesson. I thought I wanted students to work on developing their own ideas, but in reality I did not always allow the space for this to happen. The issue of sustaining mathematical activity with a high level of student involvement over a number of lessons, which I mentioned in *Section 3.2*, continued to be something I both wanted to happen and needed to work on finding ways to allow.

In taking further MEd modules I became interested in the issue of algebra and in ways of sustaining work on algebra with students. One seed of this interest was my awareness that in the year 8 lesson I saw Laurinda teach (see *Section 1.4*) the algebraic activity of the students had seemed very natural – a contrast to my own experiences of teaching. My understanding of what algebra is broadened significantly. I initially associated algebra with the use and manipulation of letters. Through readings and conversations, I began to see algebra more as a way of thinking, which always seemed to entail some sort of stepping back from whatever process you are involved with, to become aware of the process itself. I was interested in exploring what my students did when they were thinking algebraically.

In one assignment, I interviewed two pairs of high achieving students, one pair of 15-year-olds and one pair of 18-year-olds, as they worked on a problem. (The problem was in fact arithmogons – *see Section 2.3, Fig. 2.1*). This problem could be tackled algebraically. The major difference between the two pairs of students was the control with which the older ones first explored the problem numerically, until they had some sense of what was going on, and then moved effectively to an algebraic representation and solution (showing evidence of all three of the components of algebraic activity defined in *Section 4.1*). At the point where these students gained an insight into the structure of the problem which lent itself to the use of symbols, they were able to introduce letters, derive an equation, solve it and relate their solution back to the problem. The 15-year-olds, on the other hand, reached for the symbolism quickly and introduced letters, deriving a number of different equations which they tried to manipulate, but they became bogged down in the transformational work and lost contact with what their equations told them.

The 18-year-olds were operating with a much greater sense of control. There seemed to be evidence that they knew what they were looking for when they were exploring the problem initially, since they continued this exploration just until they gained the insight they needed.

This experience led to asking the question: would it be possible to create a classroom culture of 'becoming a mathematician' with 11-year-old students so that when they themselves were aged 15 they would be operating as the 18-year-olds did? Laurinda and I started work in September, 1998 on a project, funded by the Teacher Training Agency, to investigate this question with one mixed ability class that I taught.

I decided to offer the class the challenge for the year of 'becoming a mathematician'. A large part, for me, of 'becoming a mathematician' is learning about thinking algebraically and using algebra. We hoped the students would find a 'need' for algebra by working on activities which required algebraic thinking to make progress (see *Sections 1.4* and *5.4* for examples).

4.3 A need for algebra – see Brown and Coles (1999)

Alf's work on his MEd assignment (see *Section 4.2*) comparing 15- and 18-year-old students had led to our asking the question: 'would it be possible to create a classroom culture of 'becoming a mathematician' with 11-year-old students so that when they themselves were aged 15 they would be operating as the 18-year-olds did?' We started work in September, 1998 on a project, funded by the TTA, to investigate this question with one mixed ability class taught by Alf. In order to gain evidence of the students' developing algebraic awarenesses we stressed their need to write about their ideas and conjectures when doing mathematics and periodically we asked them to write about 'what I have learnt?' both in terms of mathematical content and 'becoming a mathematician'. In this section, we focus on the evidence of developing algebraic competence within one student as mathematician illustrated by his needing to use algebra.

The case of Alex: needing to use algebra

This case study, told in three stages, illustrates one 11-year-old student's developing use of the three components of algebraic thinking; generational, transformational and global meta-level (Kieran, quoted in Sutherland, 1997) over the first term in secondary school, leading to an example of the student (Alex) finding their own need for algebra. Interviews with Alex at the beginning and end of term, his exercise book, his writing on 'what I have learnt?', a half-term review, base-line entry test data and notes from observations of class lessons form the data set from which the following three incidents have been selected.

The first activity the class tackled ('1089', see *Section 5.4*) was a rich numerical problem that lasted for seven lessons. In that time the teacher Alf, was using strategies to allow students to raise many questions within the group although everyone also had a lot of practice with the processes of basic addition and subtraction.

Stage I: Algebra introduced by teacher but not used by student

Some of the questions students raised involved wanting to know why 9s fell in particular places in the calculations. No student was using algebra and Alf recognised that one way of answering these questions was to use an algebraic demonstration which he then did. Alex had not used algebra before. On being interviewed after the seven lessons he remarked that 'basically all of it in my primary school was sums' and further that ideas of proof were not used at primary school. After the algebraic demonstration 11 out of the 27 students could recreate the manipulations and 8 were able to extend the techniques to show other similar results within the problem. We did not, however, expect students to be able to reach for algebraic technique in a different context nor were we concerned that the majority of students might not be able to reproduce the original demonstration at this time. The possibility of using algebra to know why things work as they do was now around and this was our main purpose in introducing the algebra. At this stage Alex thought of thinking like a mathematician as 'you've just got to ask yourself why is it doing this?'

In the first interview, Alex was invited (by Alf the interviewer), to try numbers in a problem which he had not seen before. He quickly spotted a difference of 3 (*Fig. 4.1*):

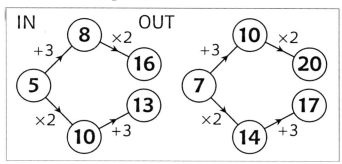

Figure 4.1: Alex's first two 'both-ways'

When Alf asked: 'You said that being a mathematician is about asking questions so what's your immediate question?' Alex replied: 'Does that happen with every number you put in?' In working on this question he tried out 'minus numbers' and decimals. It was evidently not natural to let a letter stand for any number and explore the consequences.

There is evidence of generational activity here since the pattern of 'there's always a difference of three' was spotted. But despite recognising 'why' as being a mathematical question Alex does not ask himself why in this context and consequently does not display global meta-level awareness within this problem. Algebraic symbolism was not used so there was no evidence of transformational activity either.

Stage 2: Algebra used by student in response to the teacher's question

Fig. 4.2 below is taken from a half-term review given to the class which involved some questions to explore how they were getting on with algebra and an end of half-term 'what have I learnt?'. Here, in response to the prompting in the text of the question, Alex is able to work through the problem using a general letter N (even though there is no explicit invitation to use N in the statement of the task) demonstrating some transformational skills. We believe he is able to halve $2N + 4$ to get $N + 2$ because of awarenesses formed through the numerical process of trying a few examples first. Alex is effectively using the skill of multiplying out brackets, but no algorithm for this has yet been taught. Alex recognises that the sequence of instructions always results in 2 and so uses generational activity.

3) Try out this trick with different numbers … write down anything you notice … can you prove anything about this trick?					
Think of a number ↓	1	5	100	N	The Answer always comes out as two. If
Double it ↓	2	10	200	2N	you look at the sequence most are alimaneting its
Add 4 ↓	6	14	204	2N+4	selve e.g. Think of a number, Take away the number you first thought
Halve your answer (Share by 2) ↓	3	7	102	N+2	of !!! same with Double it and Halve your answer, if there was no "add 4"
Take away the number you first thought of ↓	2	2	2	+2	it would come to 0 but there is "add 4" when that is halve it leaves
Answer	2	2	2	2	you with "2" the answer.

Figure 4.2: Question 3 from Alex's half term review

In commenting: 'most are alimaneting its selve' (we think this means 'eliminating themselves') we would intepret a global meta-level appreciation of the structure of the trick and in reaching for the N also a global meta-level awareness of the power of using symbols, although this happens in response to another's questioning.

Stage 3: Algebra needed to answer a question posed by student

In the second interview with Alex, at the end of the first term, Alf posed the same problem as in the first interview, but with different numbers. Alex tried one more example and commented: 'The one I've just done was 6 difference and the same for that one there'. As in the first two incidents, Alex displays generational activity in noticing a numerical pattern. In response to: 'What questions are around for you as you notice a pattern like that?', he replies in a similar way to before: 'Does it work for all of them?'. Previously this statement led him to try out decimals and 'minus numbers' but after one more numerical example, without speaking, this time he produced the following algebraic solution (*Fig. 4.3*).

There is certainly evidence here of transformational activity and this

feels like the 18-year-olds' interview because Alex gains control of
the process before using algebraic
skills. Even more surprisingly,
Alex returned to a numerical
problem and said: 'I know what's
making it 6 difference now, with
the N. Because the bottom way – I
can't say it. But that 7 it's going to
be more than just timesing it by 4
straight away and adding two on

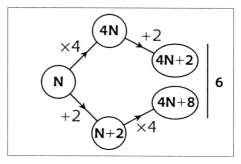

Figure 4.3: Alex's algebraic 'both-ways'

the end. Really you're timesing the 2 plus the 5 by the 4 that way. It's hard
to explain. So, that one would be 4N plus 8. So, these two cancel out each
other leaving 6 behind. So now you know every one's going to go to 6.'

Alex clearly shows evidence here of insight into the structure of the
problem, a global meta-level awareness, which, unlike at the half-term
review, is also articulated. The difference that strikes us here, compared to
the first two incidents, is that the algebra has arisen from a question of
Alex's. In recognising a pattern and asking himself 'why?' in this new
context he creates a need. His experiences over the term allow him to
answer this need with the use of a letter N to stand for a general number.
As he worked through the general case the structure of the problem was
illuminated: 'I know what's making it 6 difference now, with the N.'

In commenting on the process of his solution, Alex recognised the
power of N standing for any number: 'I should have done that first off.' Alf,
in reply during the interview, tells him that it is good to start with the
process and we would argue that Alex's need for algebra came through the
posing of his own question: 'why?' and that this came out of a pattern
spotted (generational activity) after the process of doing a few examples.

His transformational skills, in contrast to the second stage, appear less
dependent on numerical awarenesses since it is in the transforming that he
gains structural insight. It is beginning to feel as though Alex will not need
to be taught algorithmically many of the transformational skills needed in
secondary school e.g. how to multiply out brackets.

Alex had developed over 15 weeks from no experience of algebraic
thinking to using algebra to illuminate his thinking in relation to a problem.
The question we were working on for the TTA project was whether 11-year-
old students would be able to operate algebraically like the 18-year-olds in
the pilot study by the time they were 15 years old. This evidence suggested
that some of the students will be able to achieve this facility much earlier
than age 15.

4.4 Lesson account

The lesson reported on in this section, of Alf teaching his year 7 (aged 11-12) class, took place on 25/1/2001. We are struck by how the focus on algebra and algebraic thinking from the TTA and ESRC projects has meant that algebra seems to be a part of any topic that we teach. In this lesson, an algebraic proof arises from the need to consider whether a class of triangles has the same area or not. Alf would not expect everyone to be able to follow every step of the proof and in fact one girl spent the time carrying on with her own ideas and ignoring what the class was doing. The proof is, however, an example of algebra being used to do something that could not be done without it, that can motivate students to learn the skills since they see a need for them.

Pick's Theorem – year 7 (aged 11-12), 2001

We had been working on Pick's Theorem for four lessons. The problem was to find the connection between three variables; the area of a shape drawn on a dotted square grid, the number of dots inside the shape, and the number of dots on the perimeter of the shape. In *Fig. 4.4*, the area of the shape is 4 squares. There are 2 dots inside the shape and 6 on its perimeter. This was defined as an '8-dot shape'. The students had had a homework to continue working on their current question. Once the class had settled down at the beginning I invited anyone to talk about what they had been doing for homework, what questions they had been working on or any conjectures they may have formed.

After several contributions, one student said the question she had been working on was: 'How many eight-dot shapes are there?'. She had got a page of eight-dot shapes and said she had found about thirty. I asked her: 'How many do you think there are?', to which she replied: 'Infinity.' I wrote on the board: 'Kylie's conjecture: There are an infinite number of eight-dot shapes.'

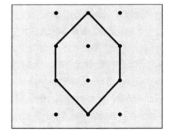

Figure 4.4: An '8-dot shape'

Anna commented that to explore this question we would need 'stronger rules' to decide if reflections and rotations were counted as the same or different. Several more students offered opinions as to how many eight-dot shapes there must be and I offered an image (*Fig. 4.5*) which I said for me is convincing that the number is infinite.

Some of the issues debated were as follows:

~ There won't be infinite shapes because it will go to a line.

~ It will eventually go through another dot.

Then Matt said: 'Won't the areas be the same?' I repeated this comment. Several other students disagreed and I said this was an issue I wanted everyone to work

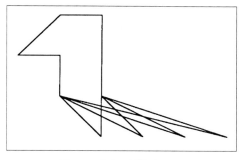

Figure 4.5: Alf's image

on. I drew four triangles on the board (*Fig. 4.6*) and told the class to copy them down and work out the four areas.

Figure 4.6: The four triangles

After giving the students some time to work on this individually (and help some copy the figures accurately) we discussed at the board ways of finding the areas. We found that each area was half a square. The homework was to continue working on a question which had arisen for them from that day's or previous discussions. There was a lot of student talk in this lesson. The dialogue below, taken from Laurinda's observation notes, needs to be read more as a series of snapshots than a continuous narrative.

Into algebra

– As you are sorting yourselves out, I want to look at what you have done for homework, as I am doing that you can look at what others have done for homework or carry on working for a few minutes.

– Obviously I wasn't in last lesson, let's start with a conversation looking at what people have been working on ... particularly conjectures or questions.

~ Kylie's conjecture, 8-dot shapes are infinite, I've been thinking about that. (He went to the board and drew *Fig 4.7*.) If you take an eight-dot shape and cut the corner off you can extend that to infinite.

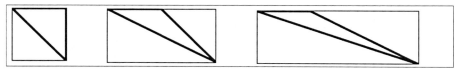

Figure 4.7: Student's drawing

~ All I can see is the bigger boxes.

~ Get rid of the boxes.

~ You know, them triangles that keep extending.

~ Can you put a limit on the number of shapes you can do?

~ The only thing I can see is different is the gap between them is one and it goes further out.

~ How is this proving Kylie wrong?

~ If it's a triangle then it will end eventually, so it can't be infinite.

~ It'll go into a straight line.

~ Keep going. Thinner and thinner and thinner.

~ I want to see if I can go there to there (points on the board) and not go through any dots.

~ Carries on going through the middle like.

~ Won't go through a dot. It's like Jim says, it'll go to a line.

~ If you do the shortest angle you could it would go to the next dot in the limit.

~ I don't agree, they'd cross over because both of the lines go to the same point.

~ As we change the shape what happens to the area? I don't think it's going to change.

– When we worked it out ...

~ It was always half.

~ Even if we go lots out ...

– Any more comments on this point?

~ Thickness of pen, microscopically thin. Would be a gap?

~ If it keeps getting thinner and thinner they must meet sometime.

~ It's not ever going to meet it's going to keep on like that.

– Why not?

~ It just won't

~ It will meet, they look like they're meeting now

~ I reckon the top bit won't meet because they're one dot away from each other

~ Even if there is something there [a gap?] it's not going to be a half

– Put your hands up if you'd like me to show you something related to this [a few]

– What was the conjecture?

Alf drew two triangles on the board, both with height 1 (*Fig 4.8*).

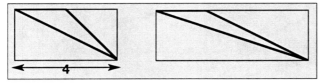

Figure 4.8: Two triangles with height 1

– I'm going to prove this for any triangle. What am I going to call this length? (Alf pointed to the triangles on the right.)

∼ A [After other suggestions it is agreed to call the length n, which Alf draws in.]

– What's the height?

∼ One.

– What's the area?

∼ What does the n stand for?

∼ Any number.

– The area of the rectangle is n times one, which equals?

Alf then led the class in chanting: $1 \times x$, $1 \times p$, $1 \times 2n$, 1/2 of 10, 1/2 of 20, 1/2 of x, 1/2 of n, 1/2 of p, 1/2 of $2x$ [one x], 1/2 of 8p [4p], 1/2 of 8z [4z]. Alf then wrote labels on the righthand triangle.

Area of triangle A = 1/2n (a half of it)
Area of triangle B = 1/2(n–1)

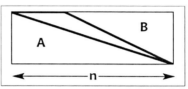

Figure 4.9: The general triangle

– If I've got area of triangle 1 and area of triangle 2

∼ Add area of 1 and 2 together and take away from rectangle

∼ Why is it n minus one?

[Kylie focuses back on the left hand triangle.]

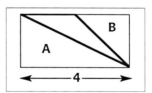

Figure 4.10: n = 4

∼ Find the area of A and B, it equals 2 add 1 and a half equals 3 and a half. Then you take it away from four, equals a half.

– All we're going to do is exactly what we've done here, but down here (pointing to general triangle).

Alf wrote on the board: 1/2n + 1/2(n–1) = 1/2n + 1/2n – 1/2
                                        = n – 1/2

– So area of bit left …

Alf wrote on the board: Area of rectangle is n
                                      Area = n – (n – 1/2) = 1/2

∼ Where's the n gone?

∼ Find the area of the rectangle and then take away.

∼ Can we stick to numbers instead?

∼ Why are we doing triangles, I thought we were doing eight-dot shapes?

~ I want to say my conjecture

– We can do this, $3 - 2\frac{1}{2}$ *etc.*

– What we seem to have proved is that the triangle always comes to **a** half. If you want to keep going try your own numbers.

~ If you take the box away, it's only a line.

~ Let's get on with it.

~ I've been so bored, I didn't listen to any of that, I did that and that and that [pointing to work in her exercise book].

(Observation notes, Laurinda, 25/01/2001)

5 EQUIVALENCE

5.1 A basis for learning

Alf: We talk a lot now about teaching strategies inviting students to compare and contrast two or more objects: I was surprised, reading the 1995 interview that I was doing something similar then but not in awareness:

> With my year 12 the first thing I asked them was to write down an equation that they could solve and one that they couldn't and then to write down the hardest one they could and the easiest one they couldn't and again wanting to get across, I don't know, I suppose at one level wanting them to get in touch with what they could and couldn't do but also trying to get across some sense of autonomous working. (Appendix 2, 1995, 2.50)

Laurinda: In 1995, you also used strategies supporting the sharing of different methods but then everyone had to do the first method that was offered, including if that method came from you, as you describe in that first interview:

> ... I asked if there was an easier way we could have started adding these numbers, perhaps we could have started pairing them. I think I mentioned the pairing idea. And someone suggested we could add the first and last and so I ... asked everyone to try and explore how many pairs for 30 ... (Appendix 2, 1995, 2.38)

Alf: It feels quite different in 1999 I'm more concerned about students talking about their methods than about everyone getting the same answer.

Laurinda: Yes, you are confidently handling a multiplicity of methods and have teaching strategies in place for discussion of them, like asking 'OK, can you explain why you did that?' (Appendix 1, 1999, 1.30). In this example of sharing the reasoning behind different methods, one of the students talks about a difference he has noticed. There's a link here between explicit use of teaching strategies for comparing and contrasting and making use of your own and the students' powers of discrimination within the classroom interaction. What's important here?

Alf: Students will only use their powers of discrimination within mathematics if there is something that interests them, if they've got a question. There was a link from the TTA research between students' need for algebra and asking of the question 'why'. When we looked at classroom incidents when students were asking their own questions that's when we found that it was when there had been a classification activity or students were being asked to compare and contrast, say, two images or examples. The ESRC research project explored teaching strategies for that.

Laurinda: What would you say, now, thinking about your teaching, are there strategies that you use which are an essential part of your practice?

Alf: I want to work with students from where they are at. One thing I know I often do is set up a situation where the students have to respond actively, *e.g., by individuals coming and doing something on the board.* What the students do gives me feedback to which I can in turn respond, and sets up a common experience from which we can start to develop language.

Laurinda: And the setting up of language, I've observed, is through comparing and contrasting, noticing what is the same and what is different. The lesson in *Section 5.3*, taken from the ESRC project, is one example of this and the lesson account in *Section 5.4* where students all do a calculation first and then differences are explored. It does feel now as though this is something that you look for in your planning. As in the ways of using boards in your classroom as spaces for the collection of information that the class shares (what we came to call 'common boards'). The teaching strategies linked to equivalence don't seem to be separable – they all contribute to the classroom culture and you now make complex decisions about how to act very quickly and flexibly contingently on what the students do. This feels quite different to the way that the findings from the TTA project were listed and yet they are accessible in this form. What do you make of that list now?

Alf: The four factors from the TTA research that we identified as being key elements in establishing a classroom culture in which students found a need for algebra were:
- giving the students a purpose for the year of 'becoming a mathematician'
- commenting on and highlighting the mathematical behaviour of students whenever it was observed – metacommenting

- the choice of activities and teaching strategies used within those activities *e.g., the use of common boards*
- an emphasis on students writing both in the act of doing mathematics and in reflection on what they have learnt.

All these strategies still feel like an important part of my practice. In my first few lessons with a year 7 group I will set up the purpose for the year in much the same way as I report in the 1999 interview:

> And I said welcome to mathematics at secondary school. I then said that one difference they might find or there might be between mathematics at secondary school and at primary school is that as well as all the skills and techniques like adding or multiplying or taking away that they will have learnt and they will continue to learn it's also about learning to become a mathematician. About becoming a mathematician and learning to think mathematically and what that means and that what I mean by that is if you're thinking mathematically then it's about noticing things about what's around you and it's about writing things down about what you notice and often what they'll be writing down will be a question about something which they've noticed – maybe they've seen a pattern and a question that mathematicians often ask is 'why?' so they might spot a pattern and think about why does that pattern work. Make a prediction maybe based on that pattern. Say why they think that pattern will continue. Did I mention anything else? I think those were the main things, some sense of noticing patterns, writing and asking questions I think were the things I focused on. (Appendix 1, 1999, 1.4)

Metacommenting on students' mathematical behaviours is something that, if anything, I think is more important now in establishing a classroom culture. 'Self-checking' and 'self-generative' activities were words I remember we used during the TTA to describe the types of activities that allowed my attention to be on the mathematical activity of the students so that I could metacomment as often as possible. I don't explicitly think about those words in my planning now, but I think I am unconsciously always looking for activities where I do not have to be the arbiter of what is right and where students do not need me always to generate the next thing to do.

The last factor from the TTA is about student writing and, again, that still feels central to what I am trying to do. Writing seems so important because to write about what I am doing I need to be able to step back from it, in much the same way as I think I have to in order to think algebraically. So, the importance I place on writing is linked to my view of all mathematics being about algebraic activity. The TTA strategies have become part of what

Evidence box 5.1: Students asking their own questions

Homework

I wonder why some numbers go in one collem and other digits go in another. I think for the numbers to go in 99099 the first number has to be number 6 or over.

All my answers come to six. Why?

I'm going to look for the pattern in the factors.

I want to see what it would come to with 4 on each side because I agree with Charley that the lowest is 15 for 3-3.

1) Can I predict what the answer is going to be

2) Is there any that won't work

I'm now going to try to do some 5 digit numbers. I'm going to use the same as I did in the 999 colum where I put middle numbers the same.

Can I start by guess what amout of answers I will guess for 6.

I think that there will be about 4 answers for the six digit ones. Because there was three answers for the 4 and 5 digit ones and I dont think there will be another 3 answers for 6. I think that there will be an answer that is 990099.

I found out that on the first set of tricks all the answers came out at the same. (Number 6) and I cant work out why.

Questions:
1. What theories are there?
2. Are there any patterns?
3. Do odd and even number make a difference?

It is through posing their own questions that students find a need for algebra, e.g. to show why something will work for any number (e.g., Evidence Box 4.1). It is through having a question that can only be answered by using algebra that students gain a sense of its power.

I do in a much less conscious way than when the report was written. It was important, I am sure, to have found that language but, as with purposes such as 'sharing responses' they did not remain conscious foci.

5.2 Using student responses (Alf)

The year (1998/9) of the TTA research project (reported in Chapter 4) was an important one for me since for the first time, with a year 7 group, I felt the class in general had maintained a positive attitude towards mathematics and were still, at the end of the year, willing to discuss and think about problems. The notions of self-generative and self-checking activities were useful ones in terms of my planning. I was becoming quite used to working with activities which were self-generative, *i.e.*, in which the students were expected to work on their own questions and develop their own ideas, within a framework (*e.g.*, of working on functions and graphs). However, I found ways of making activities self-checking were more problematic.

I had for several years worked on similar problems with classes from different years, particularly at the beginning of year 7, where I have always used '1089' (see *Section 5.4*) as the first long activity. In a first activity with a class I am particularly keen to be able to avoid getting into conversations about whether calculations the students have done are right or wrong. I want to establish, from as early on as possible, that I am not there to be the arbiter of what is true or false. One mechanism I used to avoid answering this question was to invite students to write their answers on the board and see if other people agreed. This was effective in terms of me avoiding the question but a problem was that the students did not seem particularly interested in checking each other's work.

The next year, doing the same problem with a new year 7, I got the students to write their initials next to their answers when they wrote them on the board, so that if someone else disagreed they could find the person who did the calculation and work together on where their differences were. This was slightly more effective in terms of the number of students who were checking each other's answers. I was also concerned that, as a class, we had a table of correct results which everyone could refer to, so, once a check had taken place, I invited the student who had first done the calculation to transfer their answer to a large piece of paper on the wall. One problem with this was that incorrect sums still got transferred to the paper on the wall – often because a check had not actually taken place – but there were fewer errors than in previous years.

When working on the 1089 problem in the year of the TTA research, I still invited students to come to the board and write down their answers

Evidence box 5.2: Activities

ARITHMOGONS

Place numbers in the circles such that each square is the sum of the adjacent circles.

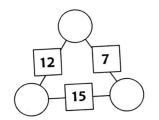

Questions that became self-generative in this activity were: can I find a solution with any numbers in the squares? (*i.e.* is there always a solution?); is there ever more than one solution? Ones that no-one can do were written up on a common board. The self-checking in this activity only involves addition and so in most cases was left up to the individual.

This activity became self-generative with the challenge to the class of getting the numbers 1 to 50 onto a class 'factor tree'. The self-checking came through using the whiteboard for students to write up their 'branches', which only got transferred onto the class display (the 'tree') if they had been checked correct by two other students. The students wrote their initials next to each branch they put on the board, then if another student disagreed they would confer. If a check proved a branch correct the student who did the checking put a tick against the initials.

FACTOR TREES

Pick any whole number (*e.g.* 15) and find all its factors: $15 \rightarrow \{1, 3, 5, 15\}$.
Ignoring the number itself, add up the remaining factors, $1+3+5=9$. Repeat the process with your new number:
$9 \rightarrow \{1, 3, 9\} \rightarrow 1+3=4$
$4 \rightarrow \{1, 2, 4\} \rightarrow 1+2=3$
$3 \rightarrow \{1, 3\} \rightarrow 1$
... until you reach 1. Display your results:

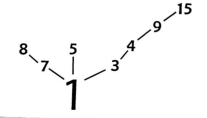

There are in fact 12 pentominoes and one self-generative challenge was whether all 12 could be tesse-lated and in what different ways each could be done? A self-checking mechanism for the minimum section of the pattern needed to generate the whole tiling was to get a student to draw their 'minimum' on the board and get the rest of the class to try and continue it on paper.

TESSELLATION

Choose a pentominoes (a shape made from five squares arranged edge-to-edge and corner-to-corner), can you find a way of arranging copies of it so that you could cover a floor (in an organised way) with no gaps.

What is the smallest section of the pattern you would need to give a floor-tiler so that they would know exactly how to continue the pattern? e.g. does this give enough information?

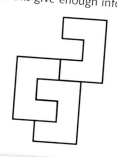

GRAPHS OF RULES

A rule might be 'times by 3', *i.e.*:

1 → 3

2 → 6

3 → 9, etc.

N → 3N

These pairs can be plotted as co-ordinates (1,3); (2,6); (3,9); etc. to make a graph.

Challenge: can you predict what the graph is going to look like from the rule?

We started this activity looking at the rule 'times by itself' and everyone plotted the graph for this. The self-generation came from students then choosing their own rules to create a table for and plot. These graphs were done on paper and pinned onto a common board so that others could (a) check they were correct and (b) look for patterns or things that are always true (*e.g.* the more you 'times' by the steeper the line).

with their initials next to it and I still had a sheet of paper on a nearby wall for checked results. However, this time, when someone had checked an answer I asked them come and write a tick next to it on the board. Only when an answer had got two ticks could the person who first wrote the calculation transfer it to the piece of paper and rub it off the whiteboard. This mechanism was (finally!) effective in terms of getting answers checked and getting an accurate table of checked results. I felt the activity was genuinely self-checking amongst the whole class, and I did not need to arbitrate.

The whiteboard and paper in the story above are examples of what, during the ESRC project [1], we came to call common boards. These were spaces where students could record aspects of what they were doing which were potentially useful for others or which could be incorporated in the work of others. Often, I would write down all the questions different students were working on during an activity. There was certainly evidence of students finding these boards helpful in terms of deciding what to do and picking up ideas from others in the class. The use of common boards seemed to be a practical way of supporting students in working on their own questions.

A teacher of English at my school came to observe a year 7 lesson, when the students were using the white board and paper and the mechanism of the 'ticks'. Partly due to her interest I started and ran a cross-curricular group of teachers, looking at teaching strategies that can be used across subjects, to promote student involvement, choice and independent thinking. A number of teachers in this group were able to adapt the mechanism of common boards to their own subject. The text below was spoken by a teacher of German, Nick Sansom, who is part of the group, during a taped conversation with Laurinda:

> Well one of the things I've done recently … it was a vocabulary building exercise we're learning about … one of the topics is what they do in their free time. And the main verbs that are used are things that they play and places that they go. So you've got 'to play' and 'to go' … And I let them with the dictionaries start looking through and finding things and using their imagination, what they play, where they go, find the words. Then they come up and write them on the board. And if another group had the same thing then they compared the spellings, see if it was right. So the end of the activity, we'd go through and select and ask if there were any problems with the spelling, and, have you got this, say, are there any differences, ask 'who's right?' then cross reference with someone else. So that worked out quite well in terms of building vocabulary …

they ended up writing up a communal list or adding bits to their own lists from the list on the board. And I think they may well have got more information than they would normally have had.

The single idea that, in the cross-curricular group, we have found has the most widespread application is that of using 'same/different' *i.e.*, getting students asking questions, or starting a discussion by presenting them with contrasting images or examples. In mathematics, a lesson start I have used is to draw the following images on the board and ask the class; 'describe what you see?' or 'what is the same and what is different about these rectangles?':

Figure 5.1: Two rectangles – what's the same? what's different?

A discussion has always ensued about the property that these rectangles have the same perimeter but different areas. Often a question such as, 'what's the biggest area we could get with this perimeter?' has come naturally from a student and has provided the motivation for a number of lesson's work.

A teacher of History, Kay Attwood, described how her views about teaching year 7 had been influenced by the work of the cross-curricular group:

> I wanted to introduce this concept of 'becoming a historian' in a wider forum and to promote an atmosphere where pupils take a far more active role in the classroom and it's less teacher led. ... trying to make sources more meaningful, that's sort of my own agenda. Our greatest problem is about helping pupils understand the questions surrounding the reliability of evidence at GCSE. And obviously my job is to start that on the first day in Year 7... get the students to actively look at sources more and reach their own conclusions more. When I talk to them about issues they can see what I'm saying but then the minute I stop they don't think for themselves. And it's trying to get them to proactively think – Who wrote the source? Why was it written? When was it written? Is there bias? Is it subjective? Objective? And to use those terms to think about reliability.

One way she had successfully used of getting students to 'proactively think' was starting a lesson by giving everyone two sources of evidence about the period they were studying. The first task was to comment about what was the same or different about these sources and the information they held. Students commented, among other things, on the distinction between one being a primary and the other a secondary source of evidence (without using those words) and a number of questions were raised that the students wanted to know the answer to about the historical period.

5.3 Same/different – see Brown and Coles (2000)

From the TTA project, specific factors were identified as supporting the development of a classroom culture in which students found a need for algebra. One of these factors was students asking and attempting to answer their own questions. Related teaching strategies included the use of common boards (see *Section 5.2*). When we looked through lesson observations to identify patterns in when and how students had asked their own questions, this had happened when students were offered contrasting examples to compare. Related strategies we came to call 'same/different'. We report below on two case studies that illustrate the complexity of relationship between the use of common boards and same/different in supporting classroom interactions.

Case study 1

At the start of a lesson the students' responses to homework are often collected on the boards. In this typical illustration, Teacher A organised the students to draw shapes with order of rotational symmetry four that they had designed for homework. At the end of the lesson the board looked like this:

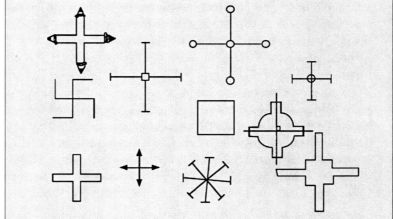

Figure 5.2: Teacher A, Observation 4, 11/11/99 – Whiteboard at end of lesson

Here is an extract of dialogue, transcribed from the videotape recording, from the beginning of the lesson:

– If you get your homework out in front of you please. What was it? Just order 4 was it? [Hands go up. 3 board pens are given out: students draw a shape and pass the pen (see *Fig. 5.2*)]. Look at the ones on the board, are they the same? They might have things that are slightly different.

~ That's a Nazi symbol.

– Given that it's Armistice Day, let's think of it as a Tibetan peace symbol because it is that as well. Comments?

~ They're all crosses in one way or another.

– All these according to you are order 4. What's similar about them?

~ 4 lines – across, down or diagonal.

~ All got 4 lines.

~ Put in a mirror – you see a reflection.

– Any that haven't got reflection? If we've got order 4 rotation, would we have lines of symmetry? . . . Is that always true?

~ All the shapes that have order 4 have 4 lines of symmetry.

~ I've found one that hasn't.

– Any on the board?

~ That one, the peace symbol.

~ If you put a mirror, you could join the lines up and it would have 4 lines of symmetry. (Observation, Teacher A, 11/99)

The collecting of homework examples provided a rich source of images for which Teacher A invited the consideration of the questions 'what is the same?' and 'what is different?'. In an interview he talked about using these questions:

> If there's more than one thing you're going to do comparisons and you're going to talk about what's the same and what's different. So, I haven't had the feeling that it's something artificial, it just seems natural to the whole sort of discussion of 'can you explain it?', 'what's going on here?', 'why do you think that?'. (Interview, Teacher A, 12/99)

The teacher's use of same/different is contingent upon the students' observations: 'All got 4 lines and . . . you see a reflection'. Teacher A mirrors back a question offering: 'If we've got order 4 rotation, would we have lines of symmetry?' to which a student makes a conjecture: 'All the shapes that have order 4 have 4 lines of symmetry', and another student gives a counterexample: 'That one, the peace symbol'.

The teacher's use of same/different to support mathematical argument is more sophisticated but nevertheless contingent upon the students' comments. In the interaction, space is created for the students to make connections. They are exploring whether things always work and if not under what conditions they hold. The teacher had not expected the lesson to go in the direction it did. His planning had been on the level of structure rather than detailed content *i.e.*, how to work with the range of homework the students would bring with them.

In the second whole day meeting of the group of teachers and researchers working on the ESRC project, having watched two contrasting videotape extracts and worked on discussing the samenesses and differences between them, we started to talk about the two teachers being 'fuelled by the kids' and worked on observing teaching strategies which supported this. A teaching 'theory', a purpose created by and for the group of people involved, had developed and allows the possibility for extending our range of practices as teachers, which then feeds back into how we think about the theory. It is also clear that the language we are developing is only useful for the group of us who went on this particular journey.

Case study 2

A second teacher, Teacher D (who was in fact Alf), worked on three problems for most of the first term of the school year, September to December, 1999. The following extracts are taken from the written work of six students from this year 7 class, who were interviewed. The six students are across the spread of achievement within the class and, as a group, indicated by standardised school tests at the start of the year, are below national averages. We were looking for students use of same/different and asking their own questions. We have edited the students' writing for spelling.

Problem 1: 1089 (see *Section 5.4*): A number trick where students are initially invited to take any three digit number where the units digit is smaller than the hundreds digit, *e.g.*, 453 and subtract from it the reversed number, 354. Reversing the answer to this calculation and adding gives 1089. What happens with four digits? five digits? The common boards are used for collecting, checking and analysing results.

Student 1:   I notice that the same numbers keep appearing in the addition part.

Student 3:   It is impossible to find any other answer than 1089. That is my opinion.

Student 4:   This didn't work. When I did the first two numbers (897 − 798 = 99) the answer was less than 100. If I do the two numbers and the answer is less than 100 it always adds up to 198.

Problem 2: Polyominoes: How many shapes can be made from squares of the same size joined edge to edge and corner to corner as the number of squares increases? The common boards were used to share methods for being organised about knowing how many there were.

Student 2:   If you are very organised you will get them all. By being organised we are rotating one square and starting from the highest point. We don't allow duplicates or if you can reflect it.

Student 5:   How many different shapes can you get with 4 squares?

Student 6:   I am being organised by putting all of the hexominoes in the right order and making sure that there are no hexominoes that are the same.

Problem 3: Functions and graphs: A game was played where a rule or function (of the form N goes to $2N + 1$) was guessed and pupils described how they had found the rule in a range of different ways on the common board. The invitation was to plot such rules as graphs and explore.

Student 1:   I have looked at various rules that got the same answers and I worked out they are in fact the same rules, just written differently. We found that $(2N + 5) \times 2$ is the same rule as $4N + 10$. So, if we had $(3N + 4) \times 5$ it is the same as $15N + 20$. But is $(N + 3)^2$ the same as $N^2 + 3^2$?

Student 2:   On the graphs there are two types of lines curved and straight. There are different rules to make either line. I have worked out both types of rules. Here are two conjectures. 'If you have $N \times N$ in the rule, it will be curved' and 'If it is straight then it doesn't have $N \times N$ in the rule'.

Student 3:   If we try different rules will the graph still work? Can we be organised in finding all the rules?

Student 4:   A question I am going to do next is 'Is it always a curved line with two n's?'

Student 5:   From what we were talking about I have learnt that not all the graphs are straight and not all the graphs are curved.

Student 6:   When it goes up in curved lines, why isn't it going up in the tables?

During the term we observed students:

• using their powers of discrimination simply to describe the problem contexts or what they notice or have learnt. For examples see Problem 1, students 1, 3, Problem 3, student 5.

• describing a purpose for what they were doing linked to specific actions. For example;

  – Problem 2, student 2, 'By being organised we are rotating one square and starting from the highest point.';

  – Problem 2, student 6, 'I am being organised by putting all of the hexominoes in the right order'.

- asking questions which were generated from the students' own use of same/different. For example;
  - Problem 2, student 5, 'How many different shapes … ?';
  - Problem 3, student 1, 'is $(N + 3)^2$ the same as $N^2 + 3^2$?';
  - Problem 3, student 2, 'Here are two conjectures.'. In the context of this classroom culture a conjecture implies something to test and therefore implicitly asks a question;
  - Problem 3, student 3, 'If we try different rules will the graph still work? Can we be organised in finding all the rules?';
  - Problem 3, student 4, 'Is it always a curved line with two n's?';
  - Problem 3, student 6, 'why isn't it going up in the tables?'.

What strikes us is how it is in the later problems that the students were using same/different to generate questions and foci for taking forward their thinking. In their written work for the first problem they were making simple observations without making a link to what they could do next.

5.4 Lesson account

The account of these lessons was written up by Alf and is in his voice. They took place with his year 7 group in 1999 and began with one of the first lessons of the year. The students were sitting in groups of four. At the beginnings of years we are aware of the greater frequency with which Alf metacomments on the mathematical behaviours of the students and the importance of the role of metacomments in establishing the classroom culture. The metacomments in this lesson account are italicised.

1089 – year 7, 1999

– I want you to write down in your books a three-figure number – does everyone know what a three figure number is? – where the first number is bigger than the last … Can someone give me their number to check you've got the idea … I'm going to pick 472. [I write it on the board].

– Did anyone else pick that? *Good, I'm just checking mine is different from all of yours.* Has everyone got one?

– Okay, underneath your number will you write the same number the other way around. So I'm going to put 274.

– All done that? Right [said slowly] I'm going to ask you to subtract your second number from your first. Now, in primary school you may have been taught different methods of how to do this. It doesn't matter what method you use and if you don't have a method then just try and find the answer in any way you can.

– Okay, I'm going to do mine on the board. I'm going to use a method but it doesn't mean any of you have to use the same method – if you have a method that works then stick to it. Well, if I add 200 to 274, I get 474. But ⌐ only want to get to 472, so I've got 2 too many. So what must my answer be? Yes, 198.

– Now swap your answer around like you did last time, so I'm going to get 891.

On the board at this point, I had written:

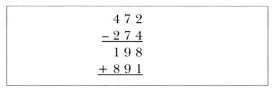

$$\begin{array}{r} 4\ 7\ 2 \\ -\ 2\ 7\ 4 \\ \hline 1\ 9\ 8 \\ +\ 8\ 9\ 1 \\ \end{array}$$

Figure 5.3: Alf's calculation written on the board

– This time I want you to add both your numbers.

~ Sir, Nina's got the same as me.

~ And Sophie's got the same as well.

– What did you all get?

~ 1089.

~ I got that as well.

~ So did I.

~ I didn't.

– What number did you start with Niki? (Niki goes through hers on the board and this time it ends up as 1089).

– *One of the things about working mathematically is that often if you start going through what you've done, you realise if you've made a mistake.*

– Shall we finish off my one to see if it works? [It comes to 1089.]

– *Okay, we've talked about how asking questions is one of the key parts of becoming a mathematician*, so what questions can we ask here?

~ Why does it work?

- There's a question before that.

~ How does it work?

– Yes, or even, does it always work? One challenge that I set you was: can you find me one that doesn't work?

A few students have answers different to 1089. I asked them to write the numbers they started with on the board, with their initials. I invited other students to check these results and, if they got a different answer, to find the person who first did the calculation and go through where there was a difference in workings.

– *This group noticed something about their answers – it proved not to be 100% correct but it's an example of what it means to think as a mathematician.*

– This group had an idea which they wrote down and tested and found it didn't work so they changed their idea, that's a great example of what it is to think mathematically.

~ I can't find one, sir.

– What do you mean?

~ They all come out 1089.

– Oh, I wonder why?

– If you notice anything about your calculations, or what you're doing, then it's really important to write it down.

~ I've tried loads and they all come out as 1089.

– Do you have any idea why?

~ No.

– Did you notice anything about the answers?

~ Yes, they all have a 9 in the middle.

– Right, that could be something to do with it – will you make sure you write that down.

If you get stuck thinking 'why?' you could go on to see if the same thing happens with 2 or 4 digit numbers.

~~~

By the fourth lesson (of seven on this activity) different students were working on different challenges (*e.g.*, some were; working on the same process with four digit numbers, sorting out how to subtract, trying to find a way of convincing themselves that three digit numbers always go to1089).

The challenge for the four digit problem was whether the students could know what the answer would be without doing the calculation (since, unlike the three digit problem, three different answers are possible). As individuals tried different starting numbers, they recorded their results on a central board, with their initials next to them, and were encouraged to check other peoples' answers. Only when answers had been checked twice (indicated by two ticks on the board) were they transferred onto a permanent display and into the students' exercise books. *Fig. 5.4*, below, is a copy of the permanent display at one stage in the lesson, and *Fig. 5.5* the whiteboard which students were using to check their answers.

10890	9999	10989
5213	9268	4333
9768	8367	6113
2761	9463	8443
	9675	7336

Figure 5.4: Checked results table for 4-digit starting numbers

– The question is – can you start predicting – can you tell me what column the number goes in?

– It's a good idea to write down a sentence saying what you're going to be working on.

– If you're not sure how to take away then that's one thing to sort out this lesson.

Some students predicted that if 4 digit numbers had two middle digits the same, then the answers would end up as 10989.

– As well as checking the ones that work *i.e.*, 4333, 6113, 8443 … make 10989 then look to make sure in the other two columns there are no numbers which are the same in the middle.

– Other people saw three was always at the end and someone checked that that was not important – so *they are noticing what they are doing and thinking mathematically.*

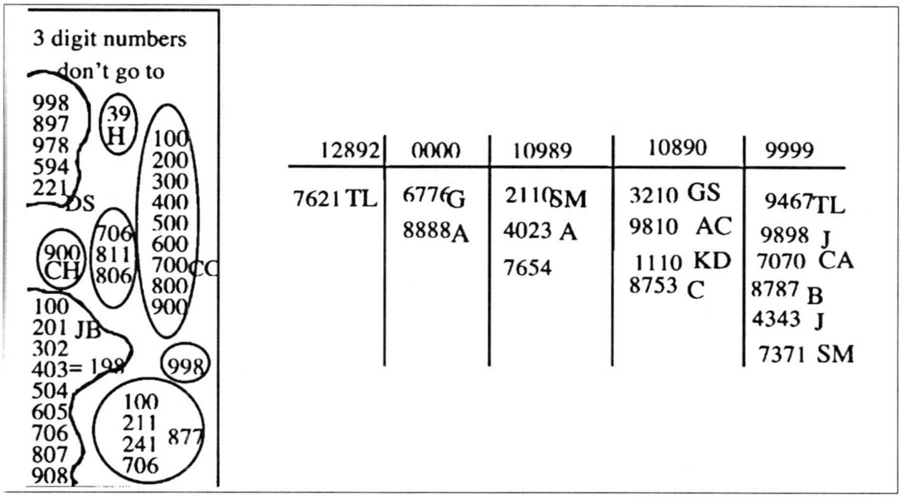

Figure 5.5: *The whiteboard students were using to check answers, 09/99*

By the sixth lesson a number of students were wondering why 9s appeared in various places in the calculation and were working on this question numerically. Everyone in the room was able to carry out the process of calculation but no one had introduced algebraic notation and it seemed appropriate that Alf demonstrate the result for three digits algebraically.

The following sequence took place (written in Alf's voice from notes by Laurinda, 9/99):

With twenty minutes to go I stopped everyone and went through the work shown here (see *Fig. 5.6*), which I introduced as a way of proving what we found out for three digit numbers. Alongside the algebra I followed step by step with a numerical example and at each stage of both the numerical and algebraic example I elicited answers for what to write from the class. I then wiped the proof off the board and set the class the challenge of reproducing it and then extending it to prove things they had found out about the problem with different numbers of digits.

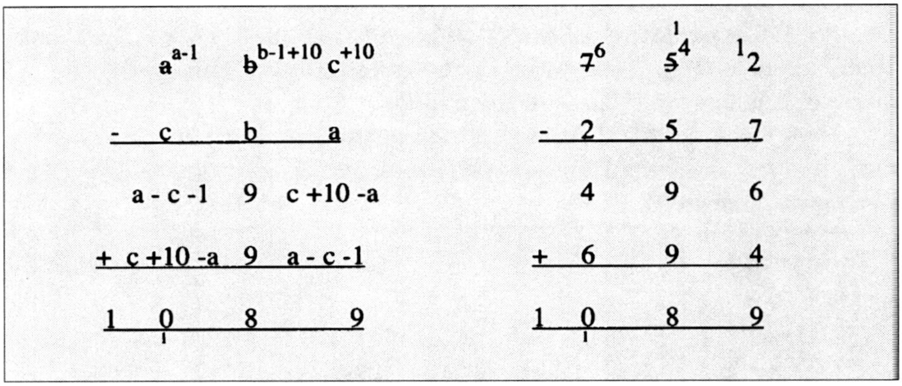

$$
\begin{array}{ccc}
a^{a-1} & b^{b-1+10} & c^{+10} \\
-\;\;\; c & b & a \\
\hline
a-c-1 & 9 & c+10-a \\
+\;c+10-a & 9 & a-c-1 \\
\hline
1\;\;\;\;0 & 8 & 9
\end{array}
\qquad
\begin{array}{ccc}
7^{6} & {}^{1}5^{4} & {}^{1}2 \\
-\;\;\; 2 & 5 & 7 \\
\hline
4 & 9 & 6 \\
+\;\;\; 6 & 9 & 4 \\
\hline
1\;\;\;\;0 & 8 & 9
\end{array}
$$

Figure 5.6 Alf's algebraic proof with numerical example

We judged that the students were comfortable enough with the process of the calculations to be able to look at that process as an object and the algebra could be said to hold that process.

There was a moment of recognition amongst the students during the demonstration when the 9 arrived in the central place of the subtraction. One student articulated 'that's where the 9 comes from'. The possibility of using algebra to know why things worked was now part of the culture of the classroom. We were not concerned at this time with how many students could perform the demonstration autonomously.

6 DEVELOPING TEACHING

6.1 On listening

Alf: In 1995 I was aware of wanting to generate classroom discussion but hadn't developed the teaching strategies necessary to do this:

> I didn't really bring them back together again but looking back, that could have been quite nice because I offered to them to write it up that evening, (Appendix 2, 1995, 2.46)

Laurinda: Simply bringing the students together again is not all there is to it, actions need to be linked to purposes. But with a year 12 class in 1995, the beginning sequence, which you quoted in the previous chapter, did have an action related to your purpose of wanting to get the students in touch with what they could and couldn't do:

> With my year 12 the first thing I asked them was to write down an equation that they could solve and one that they couldn't and then to write down the hardest one they could and the easiest one they couldn't. (Appendix 2, 1995, 2.50)

Alf: Yes, but in 1995 I had no idea what to do next, except give the students a sheet of example questions to do. That lesson beginning is an example of what I described at the time as being able to do interactive starts but without being able to sustain the journey. I think part of this 'not knowing what to do next' was that I found it hard to engage the whole class in a discussion where there would be a genuine exchange of ideas, so nothing ever really developed.

Laurinda: And that's changed by 1999, when you describe events in much more detail throughout the lesson and use many different teaching strategies [italicised below], as appropriate, to support the flow of the discussions about the problem:

> 'So, I then *asked for any comments* about that and then I've lost some of the detail around here. I think I then went back to the original person and *asked him what he thought about the second suggestion* and I think he said 'Yes, I can see that that needs to be 1 plus 2 plus 3 all the way up to 26'. So, *I wrote that.* I know there were a few other comments then at that stage and then I said 'OK, well the question was 'Can you work out how many names were said?' *so I'll give you a couple of minutes to try and see if you can work on that* and maybe see if you can work on a quick way of doing that.' (Appendix 1, 1999, 1.26)

> '... Some of them were doing things like adding the first four up and then doubling that and so *after a few minutes I stopped them and I asked for what different answers people had got* and so we got a list of about 12 different answers. *I just wrote them up* at that stage and then *I asked ... can anyone give me a reason behind any one of their answers.'* (Appendix 1, 1999, 1.28)

Alf: Supporting student discussion is something I am still very much developing in my teaching. As well as reflecting on events after the lessons I have found the access to audio and video-tape recordings of my own and others' lessons invaluable. It's also been important hearing others talking about their teaching and others' interpretations of events, on say a video-tape. As part of the ESRC project I drew a diagram to try and get across something about how the process of development has a feeling of circularity (Varela *et al*, 1995) to it:

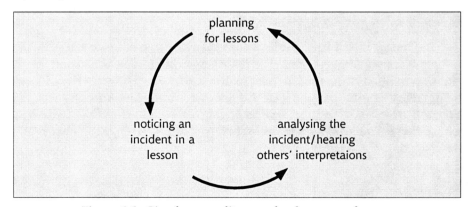

Figure 6.1: Circular causality: a cycle of events and causes

It feels as though I am engaged in an on-going process; who I am influences what I notice, which constrains what I can work on and analyse which in turn influences my planning. The cycle continues, since my planning for a lesson shapes what I will notice and so on.

Laurinda: I recognise the sense of a cycle. Watching videotape extracts with other teachers and researchers allows a broadening of perspectives and the development of new labels. One powerful label that emerged from discussion amongst the teachers [on the ESRC project] was lessons being 'fuelled by the kids'. This was part of a language of description for a perceived sameness across observed teaching strategies. Labels or purposes of this sort influence future practice through the actions of related teaching strategies, and the cycle continues.

Alf: It is possible to start anywhere on the cycle and see all subsequent events as caused by that starting point. But it's the whole process itself that seemed crucial, on the ESRC project, to the sharing of practice and identification of effective teaching strategies for us as teachers and as researchers.

Laurinda: So, where are you now? What are the incidents in lessons that are currently taking your attention?

Alf: My initial awareness of silence led to work on related teaching strategies, and one result of that was the opening up of discussion in my classroom. In reflecting on differences in those discussions, I have become interested in issues to do with listening and hearing. My sense was that the quality of listening and hearing in my classroom was a key factor in whether any learning took place. I wondered whether anything could be said about listening and hearing in a classroom. I wrote about these ideas and how they relate to my teaching as part of my MEd dissertation:

> The dictionary definitions of listening and hearing are:
>
> **hearing is**; 'the action of the faculty or sense by which sound is perceived ... the action of listening ... knowledge by being told' **listening is**; 'the action or act of listening ... to hear attentively ... to give ear to ... to pay attention to ... to make an effort to hear something' (Little *et al*, 1973)
>
> There is an overlap in these definitions in that both can be used to mean 'the action of listening'. The different aspects of the definitions of listening all share this active component; 'to give ear to' , 'to pay attention to' and in each phrase I take listening to involve an act of will or decision on the part of the listener 'to make an effort ...'. This sense of listening involving the will is echoed both in research in

psychology: 'Listening is a process that is triggered by our attention.' (Rost, 1994 p. 2) and in mathematics education: 'The act of listening ... requires a full and conscious effort to tune into the how and the what of the students' idea' (Wassermann, quoted in Nicol, 1999 p. 57).

The definitions of hearing, in contrast, I take to refer to two different phenomena. The 'faculty ... by which sound is received' (Little *et al*, 1973) seems to refer to the mechanical aspect of perceiving sound. However, the last definition of hearing in the quotation above: 'knowledge by being told' (*ibid*) does not fit with this 'mechanical' meaning. 'Knowledge by being told' implies that when I hear, something happens internally. I may be attending to whether what was said agreed or conflicted with my previous knowledge or whether what I heard extends the ideas I previously held.

There is an implied distinction here between a listening that is active but where no connection is felt with what is said and times where there is a connection made and where the hearer is changed by what they hear. I have found this distinction useful in thinking about classroom discussions but, in analysing lesson dialogue, I needed a finer grained and observable categorisation. The definitions that follow are based on Davis's (1996) notions of the evaluative, interpretive and hermeneutic listening of teachers, which I adapt for analysing the listening of students as well as teachers.

(i) Evaluative listening

If a teacher is listening in an evaluative manner then they will characteristically have a 'detached, evaluative stance' (Davis, 1996 p. 52) and they will deviate 'little from intended plans' (*ibid*). For such a teacher: student contributions are judged as either right or wrong ... listening is primarily the responsibility of the learner' (*ibid*). The teacher makes assumptions based on a supposed 'knowledge of the other's subjectivity' (*ibid*) or rather the assumption is the students have knowledge of the teacher's subjectivity – hence it is the student's responsibility to listen and learn from the unproblematic access they will thus have to the teacher's thinking.

If students or teacher are listening in an evaluative manner then they would see what others say in terms of right or wrong, and see listening as the others' responsibility. This is indicated by, for example, someone responding immediately to another's suggestion with a judgement that it is incorrect (or correct).

(2) Interpretive listening

Interpretive listening is characterised by an awareness of the 'falli-bility of the sense being made' (Davis, 1996 p. 53). If I hear someone while listening in an interpretive manner then along with whatever connection I make, or any idea that arises, or whatever meaning I take from the words, I am aware that this may not be the connection, idea or meaning the speaker intended. There is a recognition that listening requires: 'an active interpretation – a sort of reaching out rather than taking in' (*ibid*). A response might offer feedback to the speaker not by evaluating what is said but *e.g.*, by offering an inter-pretation and asking for clarification.

(3) Transformative listening

What distinguishes transformative listening from the previous category, interpretive listening, is that the interpretive listener is still 'standing back' from the speaker. There is an attempt to interpret and make sense of what the speaker says, but always from the point of view of the listener.

When I listen in a transformative mode, then as well as an awareness that what I hear may not be what the speaker intended (character-istic of the hearing of interpretive listening) I am open to the interrogation of assumptions I am making, *e.g.*, that allow me to believe communication is possible at all.

I have again drawn on Davis's (1996) categories of listening. He defines his third form of listening (which he labelled 'hermeneutic') as:

... an imaginative participation in the formation and the transforma-tion of experience through an ongoing interrogation of the taken-for-granted and the prejudices that frame these perceptions and actions. (Davis, 1996 p. 53)

The notion of the 'transformation of experience' links this form of listening to traditions of Buddhist mindfulness, in which knowledge is seen as 'equivocal' and 'open to question or revision' (Claxton, 1997 p. 219).

Evidence of transformative listening and mindfulness in a classroom includes a willingness to alter ideas in a discussion, to engage in dialogue, to entertain other points of view, and hold them as valid, independent of whether they are accepted or not. If a student makes a connection to a previous piece of work or links something that has

been said before, this would indicate the transformation of experience, the re-structuring of categories. Similarly, if a student creates a
new categorisation, this indicates a mindful attention to what is
happening: the seeing of 'a new world' (Thera, 1996 p. 32). This
sense of re-structuring previous categories or ideas, seeing a 'new
world' is indicative of learning. (Coles, 2000 p. 13-19)

6.2 Being a teacher (Alf)

I would now say of myself that I am a teacher. It took some years before I
felt able to use that word in self-description. What made the difference I
think was learning to listen and learning to hear. I began teaching with
ideals of promoting student autonomy, which have returned for me recently
as something I think about. The image of how I would like to be in a classroom, that I wrote in 1995, is still something I find relevant:

> 'I see myself being matter-of-fact, avoiding condescension and blame
> and even praise; capturing attention and emotion and directing this
> into students' own awarenesses. I would like my students to feel that
> Maths was something that connected with them, that its roots lay in
> their own intuitions, and that it can be tackled independently. I
> hoped the study of Maths would help students to realise the poten-
> tial and power of their minds.' (Diary, Alf, 07/95)

At the time I wrote this I had few strategies for trying to promote
independent thinking, 'autonomy' was too big a word, and too far removed
from my day-to-day practice to have an influence on what I did. As I look
back now over the last five years I think I needed to find labels like 'silence'
or 'metacommenting' or 'it's not the answer that's wrong' to think about,
which were more grounded in action. A purpose, like 'it's not the answer
that's wrong' is directly linked to what I may choose to do in a lesson (*e.g.,*
see *Section 3.3, p. 48*).

The purpose I set myself a few years after beginning teaching of: 'don't
comment, metacomment!' seemed an important one in forcing me to listen
to students and in allowing more space for students to work on their own
understandings and ideas. Similarly the notion that the purpose for the
year for students is to learn about 'becoming a mathematician' helped shift
my focus onto the sense students were making of situations and to become
more attuned to what approaches they were or were not adopting. I now
find I am aware, for the first time in my teaching, of when students say
things which are the inverse or converse of what another has said. Inverse
and converse are, for me, important aspects of thinking mathematically. I

am sure that students from when I started teaching have used inverses or converses, but I am only now noticing them. Each year with my year 7 classes particularly, as I become more aware of, and comment more on, aspects of thinking mathematically it feels as though a more sophisticated mathematical classroom culture develops, and I get more surprised by what students are able to do (*e.g.*, see *evidence box 4.1*, the collage of students' work *pp. 62-3*).

The idea of autonomy has translated into 'students asking and working on their own questions' which, stated like this, is linked closely enough to things I might observe and do in my classroom to feel like I can explicitly work on getting it to happen.

In preparation for lessons I do not try to predict exactly what will take place. I expect to act in response to what students say or do. Through having worked on purposes such as 'sharing responses' I have a range of strategies for organising students (*e.g.*, a student describing their idea at the board, students working in pairs for a few minutes to identify questions, writing on the board the list of all the questions students are working on) and I do not always feel I need to have thought about these beforehand. Working on the mathematics of the problem or activity continues to be important and I still need to have planned to conviction how I am going to begin a lesson. The more I work on a problem over time the more it feels I am able to hear and be flexible enough to respond to what students say so that the lesson structure can support them in working on the issues that they are finding interesting or problematic.

I am also a researcher. At the heart of this statement is the discipline of trying to separate my judgements about what I see from detailed description and, since this is never ultimately possible, then trying to become aware of what I am bringing to my observations – where I am coming from that means I notice what I do. I see the research process as a questioning of assumptions in an attempt to see more detail in the classrooms (my own and other's) that I observe. Evidence for me is just these observations. Gathering evidence requires me to notice something and the more I notice the more I have to work with.

I am about to begin the academic year 2001/2 as acting Head of Department in the school I have taught in for the last five years. This is a responsibility that in some way seems to require a combination of teacher and researcher. I am clear that I do not want to impose my own ideals or beliefs on other staff. What I do want to happen is that, as a group, we work on developing our awareness of what happens in our classrooms and that we question our habitual behaviours and responses. I want to promote discussion of lessons and teaching strategies and hope that we will identify purposes so that a meaningful exchange of ideas can take place, independent of what our classrooms are like. My own development as a teacher has given me the confidence to resist the lure of quick fixes or solutions to teaching mathematics (be they schemes of work, text-books, or training courses) that I am told 'work'. The fundamental issue is educating our awareness of the process of teaching and learning and I believe this is best done in our own classrooms, with colleagues who understand the context in which we work.

6.3 Developing listening skills – see Coles (2001)

The PME paper in 2001 was written by Alf and is extracted from his MEd dissertation. Alf analyses extracts from transcripts taken from videotapes of one teacher (Teacher A from the ESRC project), telling the story of that teacher's developing use of listening in his classroom. The theoretical frame of evaluative, interpretive and transformative listening through which Alf analyses the transcripts was discussed at the end of Section 6.1. Alf then moves on to a consideration of teaching strategies used.

Listening: a case study of teacher change

Methods used for this case study

There were four teachers on the ESRC project who were videotaped in each of the six half-terms that make up an academic year. The camera was fixed at the back of the classroom - focused on the board but with around half the students in view. The data for this study is taken entirely from the video-tapes of one teacher, Teacher A (TA). I was looking at times during the lesson of whole class discussion, *i.e.*, when there was a single conversation occurring in the room. I initially watched the videotapes and noted, at 5 second intervals, whether a student or the teacher was speaking. This record helped me identify times when students responded directly to each other or when there was significant interaction between teacher and students. I then transcribed (in the three transcripts below the numbering of students in each transcript is done independently) those sections of dialogue from the video recording. I chose Teacher A for the study because, of the four teachers on the project, there was the clearest evidence of a change in listening on the videotapes of his lessons.

Transcript 1:

TA: Any comments about those three numbers? [The numbers referred to are: 92 101, 29 810, 54 321.]

S1:   They all have two in them.

TA: They all have two in them [pause] they do [pause] anything else?

S2:   They all have one in them.

TA: They do. [Two more students offer suggestions, which Teacher A responds to.]

TA: Now remember what we were saying ... when we were looking at four digits we were comparing the first and the last, we were comparing the two middle ones. What can you tell me about the first and the last with those ones ... what can you tell me about the first and the last?

S3:   [unclear]

TA: Thank you S3: nine is bigger than one, two is bigger than zero, five is bigger than one. (Teacher A, 09/99)

Analysis 1

The dialogue in Transcript 1 shows evidence of evaluative listening. After the comments of both S1 and S2, Teacher A says 'they do' thus evaluating and confirming the students' contributions. S3's comment is greeted with a 'thank you' which the other comments were not, suggesting to me that this is the comment that the teacher wanted (although the comment is unclear, from Teacher A's response, I interpret S3 as saying something about the first and last digits of the three numbers under consideration). Further evidence for the teacher having a pre-given idea of what he wanted the students to say is that having started with the general question: 'Any comments about those three numbers', Teacher A then asks: 'what can you tell me about the first and the last?' Having started with an open question, since the students were not offering what was wanted, the teacher directs their attention to a specific aspect of the problem.

It seems possible here to pick out sentences and analyse them using the categories of listening. However, in viewing more videotapes this rapidly became problematic. In looking at transcripts of sections of dialogue to decide what type of listening was being displayed I needed, in most cases, to take into account the wider context of what was happening in the lesson. For example, in a different lesson a student said to his neighbour: 'You are wrong'. On the surface this seems typical of evaluative listening. However, if this comment was the start of an interaction in which the students began to explore their differences, the listening would be interpretive or transformative. It therefore made more sense to characterise whole lessons or sections of lessons as evaluative, interpretive or transformative.

In fact, when I analysed longer sections of the lesson transcribed above the listening was more interpretive. In general, Teacher A does not evaluate the students' contributions as right or wrong. However, the task for the students is to fit their comments and suggestions to the teacher's plan. Teacher A interprets the students' comments and gives feedback in relation to the idea he has chosen to focus upon.

Transcript 2:

S6:   It's got four sides.
TA: It's got four sides, okay, very good, anything else?
S7:   It's got four equal angles.
TA: Four equal angles, yes.
S1:   It's got six lines of symmetry.
TA: Six lines of symmetry, right, we're talking symmetry. Where's your lines of symmetry then?

. . .

S1:   Across the right hand top corner to the bottom left hand corner.
TA: This is a line of symmetry? [TA holds up a ruler along a diagonal of the rectangle] [pause] he's unsure. Who thinks it's a line of symmetry? Hands up [pause] a couple of you. [pause] Who thinks it's not a line of symmetry? [lots of hands go up] Oooh, okay, S3, convince those that think it is why is it not a line of symmetry do you think?
S3:   You can only have diagonals in a square.
TA: Oh right, okay.
S4:   Or a circle.
TA: Why is that one not a line of symmetry though? S5.
S5:   Well, if you get like a A4 paper, that's a rectangle, you can fold it diagonally so that it goes all [unclear] (Teacher A, 03/00)

Analysis 2

I believe the listening in Transcript 2 moves from interpretive to transformative. A student makes a suggestion: 'It's got six lines of symmetry', which is dealt with in a different manner to the ones just before. Rather than continuing the interpretive listening pattern of repeating each student's contribution and asking for other comments, Teacher A says: 'Where's your lines of symmetry then?' The teacher cannot know where S1's lines of symmetry are, hence he is genuinely involved in making meaning of the comment.

Teacher A then asks for opinion from the rest of the class: 'Who thinks it's a line of symmetry? Hands up'. After S5's comment, Teacher A gets an A4 piece of paper and starts folding it in the ways S5 and then other students suggest. The teacher responds directly to suggestions from students. The task for the class (in this case deciding what is a line of symmetry and how many there are on a rectangle) emerges from the interaction of students and teacher. I read Teacher A's comment at the start of the transcript: 'right, we're talking symmetry', which was said with a slightly higher tone of voice, as further evidence that he had not anticipated dealing with issues of symmetry. There is a feel of collaboration and participation in the dialogue – characteristic of transformative listening.

Transcript 3:

TA: Excellent. Oh, lovely. Well done. [Students applaud.] So, 3 times 4 is 12, 2 times 6 is twelve, 1 times 12 is twelve and a half times 24 is also 12.
S3:   And a quarter times 48 is twelve.
TA: And a quarter times 48 . . .
S3:   And an eighth times . . .
S4:   Three quarters.
TA: And an eighth times . . .
S: I'm not saying.
S: You can actually go on.
TA: . . . We could carry on forever couldn't we?
Ss: What about 100? How could you draw it though?
TA: Well, it would be a sixth of a unit. Very small.
S: If you drew it really big so one square was 6.
S: Sir, what would just a straight line be? (Teacher A, 03/00)

Analysis 3

The participatory nature of discussion is even more evident in Transcript 3 (taken from later in the same lesson as Transcript 2) in which the listening is also transformative. The teacher here is not running the discussion (*e.g.*, by posing questions for the students to respond to). It is the students who are asking questions: 'What about 100?', 'What would just a straight line be?'. Students are now talking directly to each other and extending each other's ideas *e.g.*, 'S3: And a quarter times 48 is twelve'.

The transcripts provide evidence that there was a significant change in the listening in Teacher A's classroom. The listening in videotapes of lessons up to Transcript 2 was interpretive or evaluative and in all later videotaped discussion the listening was transformative, so the change appears to have been a lasting one.

Teaching strategies for slowing down and opening up discussion

It is striking that there are a number of teaching strategies in evidence in Transcript 2 (and later discussions) that were not being used in Transcript 1.

These strategies include:

- the teacher asking a question they do not know the answer to. Teacher A says: 'Where's your line of symmetry then?' Having made this comment there is immediately the possibility for other students to engage with S1 in dialogue.
- responding to students' suggestions. There is evidence of this particularly in the sequence when Teacher A gets a piece of paper and starts folding it.
- asking for feedback from the whole class. Teacher A asks for 'Hands up' in response to the question 'Who thinks it's a line of symmetry then?'. Feedback from this response allows the teacher to use the next strategy.
- asking a student to explain their idea to the class.

These strategies can all be seen as 'slowing down and opening up discussion'. They are strategies that encourage and allow different students to engage in dialogue with each other. In Transcripts 2 and 3 over a quarter of the class speak in a period of a few minutes. Another way of characterising the teaching strategies is that they all depend on the teacher's contingency upon the responses of the students. It is important to note that this does not imply the teacher will do anything the students suggest, but only that students' voices can be heard and can play a part in the creation of the lesson focus.

There is evidence from other teacher's lessons on the project of the teaching strategies above being used during times of transformative listening.

It is striking that in Transcript 3 it is not the teacher who is 'asking a question they do not know the answer to', or 'responding to students' suggestions', but the students themselves. It seems that students are taking over some of the roles in discussion previously performed by the teacher. A culture of transformative listening is becoming established in the classroom. In Transcript 3, for the first time on any of Teacher A's videotapes, students raised their own questions, which they could work on, related to the mathematical activity.

6.4 Lesson account

During the ESRC project we worked on how to present lesson descriptions which might, unlike the **Mathematics Teaching** articles which had left Alf dissatisfied in his early years of teaching, support the decision-making of teachers when working with their own students. What follows is one example of our current style where there is a commentary in italics from the teacher which aims to give some access to the teaching strategies they use in many different situations. Other lesson write-ups can be found on www.mathsfilms.co.uk. This is the thinking that informs their practice.

We think of such lesson write-ups as 'Middle Points' as opposed to Starting Points'. The initial activity could lead to far more complex mathematics, or, the problem could be unpacked to allow a focus on basic skills and concepts such as directed numbers, say. The activities are rich and a sequence of lessons might run for some weeks of lesson time. During this time the students are asking and answering their own questions whilst spending time as a class discussing issues that arise. Algebraic activity is always present because for us this is the essence of mathematical activity and students continually meet algebra through meaningful contexts in which they have a need to express and explore their ideas. In this sense algebra is a natural part of learning mathematics, not a separate and problematic aspect of it. The lesson was with a year 7 class and was taught by Alf.

Functions and Graphs – year 7 (aged 11-12), 2001

In the first lesson I introduce algebraic notation as a way of labelling a rule that students are already using. The motivation to look at graphing functions will come when we have a disagreement about rules.

In silence, I write $3 \rightarrow 4$ on the board. Underneath, I write $5 \rightarrow 8$, and $9 \rightarrow$. I hold up the board pen. Several students put hands up. I nod to one, motioning for them to take the pen. They come to the board, take the pen and write 10. In a different colour, I draw ⊗ and take back the student's pen. The student returns to their seat. More hands go up – I nod to a different student who leaves their seat and writes 12 on the board. Again, I draw ⊗ and invite a third student to come to the board. They write 16. I draw ☺ and point to the space beneath the 9 (*i.e.*, indicating that I want the student to give the next starting number). They write 6, give me the pen and return to their seat.

By inviting the student to give the next start number, I hope to provoke an awareness that there is no pattern in these numbers – *i.e.*, we can put anything as an input.

I hold up the pen, and invite another student to the board. This pattern is repeated, with different students writing an answer and, if it gets a smiley face, writing a new starting number.

As more numbers are written and answered, I comment 'If anyone wants to write something that might help the others, they can use the black pen on my desk and write on the right hand side of the board'. A student writes: x 2 – 2 with the black pen, another writes: x 2 – 1. A third student writes: add the number to itself and take two. As they are doing this the game with the numbers continues.

Inviting students to write something that might help others is a strategy for supporting students in the group who may not know what is going on!

After some time, when it seems as though most of the class could write answers, I take the pen from a student and write N as the next starting 'number'. I do not comment on what they write and keep writing N as the starting number until no on else wants to respond.

I am aiming here to get as many different versions of the rule as possible, this can then provoke a need to work at whether the rules are the same or different.

At this stage the board was as below:

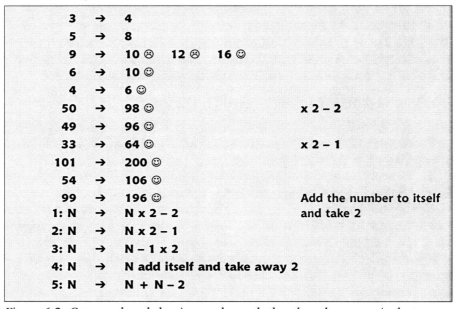

Figure 6.2: Common board showing work on whether the rules are equivalent

If there had not been differences in the rules that students wrote I would have started another game with a different rule.

Having generated a difference in students' responses to a task, it is now possible to work with those differences i.e., it is possible to ask questions or invite students to ask questions.

– What do these mean? [Pointing to the rules.]

~ The first one says times your number by two and take away two.

– Can you show what you mean with one of the answers?

~ Like the four. Four times two is eight and eight take two is six.

~ On my one [x 2 – 1] I meant take away one first.

– What do you mean?

~ Four take away one is three, and three times two is six.

~ That's the same as the third rule.

~ Isn't that the same as times two and take away one?

This question (are the rules the same or are they different?) may not come explicitly from the students, in which case I might offer the question: 'Which of these rules are the same? Will these always be the same or always be different?'

Possibilities

We might write out the rules and try substituting 10 numbers into each of them. There is a chance to focus on the conventions of writing algebra as the rules are re-written *e.g.*, 'Mathematicians tend to write 2N instead of N × 2 but it means the same thing', or, 'You need brackets around the N − 1 if you want to show that needs to be done first'.

I might play several more function games, introducing the idea of putting the right hand number first and writing the arrow to the left.

At some stage in this series of lessons, I will move the task onto drawing graphs of rules. This will happen when, having played a function game, we again end up with differences in the rules the students write. This can be particularly rich if one of the rules is quadratic (e.g., I have played a game with the rule double take five, one student writes the rule as 2N − 5, and another writes NN − 5).

– One way that mathematicians can tell if rules are the same or not is to draw graphs of them. I'm going to show you how to draw a graph of a rule. I want everyone to copy this down exactly as I do it, because every time you draw a graph I want you to do it in exactly the same way.

I write the rule on the board (in the illustrations below, the rule is 2N − 5) and underneath write 5 → , 4 → , down to −5 → . Getting answers from the class, I fill in this table and next to each line write the pair of numbers as co-ordinates. Everyone then copies a set of axes, which I draw on the board. Students come to the board to plot the co-ordinates from the table, ticking them off as they go. Students copy the points onto their axes.

Some discussion is necessary with the negative co-ordinates. I do not labour the point about rules for multiplying negatives – I see part of this activity as setting up a situation in which students can work at making their own sense of multiplying negatives, whilst their attention is on something else (i.e., drawing the graphs). Linear rules introduce an element of self-checking – if the points do not lie in a straight line then something has gone wrong.

~ All the points are in a straight line.

– If you agree with that, you can join up the points using a ruler.

– Now, in your books, do a table for our other rules (*e.g.*, (N − 5) × 2 or whatever else they have come up with) and plot the points on the same set of axes. Are the rules the same or are they different?

The difference in the rules provides a motivation to plot the graphs. If there is a quadratic rule, this will need to be done altogether in order to discuss issues such as negative two times negative one.

After plotting the first rule, the board might be as below:

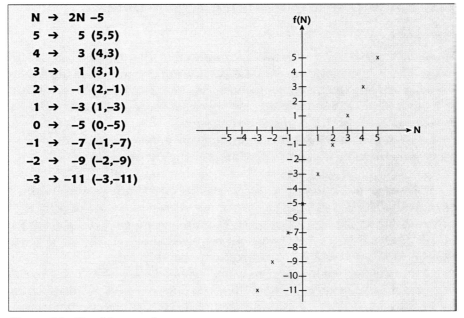

Figure 6.3: Graphing a rule

Possibilities

Once students have begun the process of drawing graphs from rules, many possible challenges are possible. If encouraged, the students themselves will come up with questions they want to explore.

– What other rules will give the same line as this one?

– If I give you any rule, can you predict what the graph will look like without having to plot it?

– Can you be organised in the rules you try?

– Given two rules, are there any numbers that will give the same output?

– Which rules give straight lines, which rules give curved lines?

One way of structuring this work is to get students to draw their graphs on paper (not writing the table of values but including the rule, written large), which they pin up on a display board. At some stage, one or two students can be given the task of organising the graphs. When organised in some way, students can be invited to gather around the display board.

– These graphs have all been organised by a student. What does anyone else see or notice?

Categorising the graphs students have drawn and discussing classifications is a strategy for provoking questions, e.g., someone might notice all the curved graphs have NN or N2 in them, question: will this always be true?

Extension

– If I have a rule $N \rightarrow aN + b$, can you describe what the graph will look like? What do a and b tell you?

– If I have a rule $N \rightarrow aN^2 + bN + c$, can you describe what the graph will look like? What do a, b and c tell you?

Mathematical Aims

- Reading algebra
- Substituting into formulae
- Plotting co-ordinates
- Drawing axes
- Arithmetic with negatives
- Gradients and intercepts of straight line graphs
- Recognising when a graph will be straight or not.

7 END NOTE

Laurinda: So, what now? We still work together and lately have started work with support from NCETM on a project related to the work of Caleb Gattegno, an influence on both of us. The story continues. I remember when we first started working together that Dick Tahta read 'The Story of Silence' and suggested that we could call the paper some other title not including the word story. He felt that the word would devalue the paper in some people's eyes as if we were writing fiction. Feels like story is a more technical word for us?

Alf: And also it feels like the academic world has moved on in this respect and 'narrative' approaches to all sorts of research are now valued. A theme of this book has been the importance of separating 'accounts of' from 'accounts for' (to use John Mason's language *e.g.*, 1996). In other words, separating description from analysis. This is complex because part of our shared theoretical belief is that there is no such thing as a pure (objective) description – all data has involved some interpretation already (*e.g.*, minimally, in noticing one particular aspect from among countless others in the moment in which it occurred). But I remember we worked hard to remove any words from 'Sarah's Story' (Account 7, p. 46) that could be seen as interpretations. The distinction I am trying to draw here is between 'descriptive' statements that are potentially verifiable by another and 'interpretations' that are not. For example the word 'surprise' was in there originally to describe Sarah's response at one stage – and we agreed that actually this was not something that could be verified. In fact I think there is one 'interpretative' statement still in the story: 'Sarah quickly replied '16' having had experience of writing functions in this way'. It reads to me now as though we are trying to explain Sarah's quick response – it is simply unverifiable whether her previous experience of writing functions was relevant to her quick response. So, story is a technical word, as you say, and I think is about beginning with something that can be observed and agreed upon before doing anything else (*e.g.*, analysing or interpreting). And this seems to me a description of a very familiar way of working we both use in any teaching context.

Laurinda: That's useful – 'what is it possible to agree on?' feels like a mechanism we use and is a way of working in mathematics classrooms. Dick, again, was instrumental in me being aware of the power of groups sharing what they saw. In the early days of my teaching, this was in relation to working with films such as 'Dance Squared' and 'Notes on a Triangle' where the first task after viewing the film was to try to recreate it – talking it through – different people having seen different aspects. Staying with the detail is another aspect of 'account of' and it is this process of looking and seeing more that, I suppose, mean that we will carry on looking ... Thinking about Dick reminds me that I want to acknowledge Dick's presence in this book – reading our writing – suggesting edits from an early draft of the book even and being available for those three-way generative conversations around his kitchen table drinking coffee that we enjoyed so much. It is still unbelievable to me that he died in December 2006.

8 APPENDICES

APPENDIX 1

9/9/99 – 7.00pm on evening of the lesson (Contributions are numbered 1.1, 1.2, 1.3 etc for reference purposes.)

1.1   *LB:* OK. You've just had all sorts of first lessons. What I want you to try to do is to imagine yourself walking through the door into a lesson and just start talking to describe what happens. I might at any stage ask questions or interrupt or ask reasons why or whatever so it'll feel a bit more like a conversation than you just talking but try to get as much detail as possible without the interpretation. Just, what did you do, what happened. So, which lesson are you thinking about talking about?

1.2   *AC:* Well, I was assuming it would be year 7. I'd just had Ian, my further mathematician, the lesson before so I managed to pack him off just about before the others arrived and then I was just in the corridor and some people were coming up wondering where C11 was, so I told them to line up outside the door. They had obviously been to different lessons because they were coming in in dribs and drabs. So, about eight of them had lined up and so I got three boys to go in and sit down, and I asked them to go and sit, the first three people to go in and sit at a table. A few more people started arriving and I got the next four or so to go and sit at a table.

1.3   *LB:* Had you prepared that beforehand? You knew that that's what you were going to do? Or did that just happen?

1.4   *AC:* Normally I suppose I would have waited for them all to come in and then I'd bring them in in groups but yeah I'd known that I was going to bring them in in table numbers and tell them which table to sit at. So, as they came in, there were about 12 people in and there was noone outside so I told them to get their bags on the floor and their books out – uhh – pencil-cases out. Then a few more people started arriving so I told them to wait outside then I went out and I told four or five to go and sit at the table and that way the room filled up and we then had 25 students.

I said to all of them I'd like them to get their bags on the floor and get their pencil-cases out and there was one girl at the back who'd come in last who was faffing and so I waited for her to get her bag on the floor and get her book out and then get her pencil-case out and that all eventually happened. Just before then someone had asked if I was Mr. Coles and I hadn't responded to that [both laugh] yeah and I think someone else had asked me something which I also hadn't responded to. So they were all there and so I was standing at the front and then they all went quiet and so I said 'Welcome' and I said my name was Mr. Coles and I wrote that on the board and I said all their maths lessons would be in C11 this room.

And I said welcome to mathematics at secondary school. I then said that one difference they might find or there might be between mathematics at secondary school and at primary school is that as well as all the skills and techniques like adding or multiplying or taking away that they will have learnt and they will continue to learn it's also about learning to become a mathematician. About becoming a mathematician and learning to think mathematically and what that means and that what I mean by that is if you're thinking mathematically then it's about noticing things about what's around you and it's about writing things down about what you notice and often what they'll be writing down will be a question about something which they've noticed – maybe they've seen a pattern and a question that mathematicians often ask is 'why?' so they might spot a pattern and think about why does that pattern work. Make a prediction maybe based on that pattern. Say why they think that pattern will continue. Did I mention anything else? I think those were the main things, some sense of noticing patterns, writing and asking questions I think were the things I focused on.

1.5 *LB:* What was going on in the group at this stage?

1.6 *AC:* They were all just listening. Ah, yes, and then I said and we're going to be working as a group mathematically and one of the things that that means is that 'there isn't a sense in which there's a right or wrong answer' because it's about learning and you often learn by making mistakes so it's not about getting things right or wrong and if we're going to be working mathematically as a group it's about helping each other and helping each other understand what's going on and also if we're going to be working as a group then we need to know each other's names. [Both laugh.] So I then asked them to put their hands up if they knew the names of everyone in the room and about three people did.

So, I said, right, good, well one thing I want to work on this lesson and I'm not sure we'll get there this lesson but one challenge you can set yourself is to leave this room knowing everybody's name. I then asked the first person on the first table to stand up and say, I'm ... and she stood up and said 'I'm Josie' with a theatrical gesture inward and the next girl stood up and I said I'd like you to stand up and I'd like you to say 'You're ...' and then say 'I'm ...' and she stood up and said 'You're Josie and I'm Laura'.

1.7   *LB:* So you asked them to stand up or they stood up themselves?

1.8   *AC:* No, I asked them to stand up. The next one stood up and said 'You're Josie, you're Laura and I'm Michaela.' and this was producing some sort of subdued giggles between these girls I think perhaps because of these sort of gestures but that sort of didn't last very long and the fourth one stood up and said 'You're Josie, you're Laura and you're Michaela, I'm Sophie.'

1.9   *LB:* Without a gesture or with a gesture?

1.10   *AC:* With a gesture still. So they all four of those gestured quite strongly.

1.11   *LB:* Were they on a table together?

1.12   AC: Mmm. And then once it moved to the next table that sort of went away a bit, then stopped.

And so we get to about six or seven people, we get the two tables so there were seven people and then I said, those people who think they've got it easy now because they've done their names, I didn't say we were going to stop after one round and I also said and you could think about 'how many names will be said by everybody?' if you finish. Then we carried on. The first four girls had no problems the next group of three boys had no problem and there were two more boys who did it OK. Two more girls, no problem. Two more girls, one of them stumbling a bit but OK, another girl fine, next two girls fine. We were then up to about 14 or 15 and the first boy who was still OK but still had quite a few difficulties and the next one was OK and there was one, probably the weakest one, about the 18th probably, but he struggled through a bit but he still had bits of the pattern. The next boy was fine, the next two girls no problem at all, next three boys did it pretty much straight off. That felt different, it had always felt like much more support had been needed in previous years. Then I did it and got everyone which produced a round of applause which was rather bizarre as well.

1.13 *LB:* What happened then?

1.14 *AC:* Then I said, does anyone else think they could do that? And quite a few hands went up so I got one person to do it and they were quite energised about that and they were zipping around quickly the ones who were doing it. I then asked if anyone couldn't do it. A few hands went up. So then I asked if the people who could do it had anything they could offer about how they could do it. But they didn't really have anything to say about that. So then we carried on a bit. I asked a few more people to do it and at this stage there were lots of hands up so I said OK, if you think you can do it, do it to your neighbour and while you're doing that I'm going to hand out your new maths books. There were lots of hands up and lots of people who obviously wanted to do it so that was just quite nice. So, I handed out the maths book.

1.15 *LB:* So, you would say that you hadn't planned this at this stage? (AC: Yeah.) It's feeling as though you're just adapting. Even in last year's tape it would have felt more like you'd just done that bit and go on to the next thing. This feels very much like these are things that you are, it's allowing you to change, 'cos you were aware that they were doing it better and because they were doing it better then there's more energy to carry it on so you keep going and you need something to deal with all those hands so the pairing happens. So this must mean at some level you've got quite a lot of embodied strategies to cope with that, to allow you to just 'pair'. That feels like probably the first time we've had evidence of that in the first lesson. I'm not sure about that, when we listen to them all one day, then ... yes, that's nice!

1.16 *AC:* It did feel nice because there was the kind of energy I recognise from last year where people wanted to speak and would have been upset if they weren't allowed to speak. But that sort of went away and then people got into writing their names on their books and so they were talking to each other and as I handed out their books and wrote up what I wanted them to write on the front of their books then they sort of got into doing that and it felt like that was OK and they'd said what they'd wanted to say. My sense was that they had been doing that and that they had been genuinely learning names with each other. So that felt quite good. So then I wrote up my name and the room and tutor group and what they had to write on the front and then I said 'something I haven't mentioned is about the rules of the classroom but I'm sure lots of teachers have been talking about that' and some girl mouthed 'all of them'. So I said that they'd been in schools for 6 years and they knew what sensible rules of the classroom were about and I was sure they were mature enough to understand what was expected of them and it would become apparent if they were doing things I wasn't happy with and that one thing I had noticed was that some people were calling out without their hands up and that was a reference back to the beginning of the lesson and so I said that that was one thing, because if we were going to have discussions that was going to make it very difficult. So, that was one thing that I didn't want them to do.

1.17 *LB:* Do they know they have to sit in the same places next time?

1.18 *AC:* No. I didn't say that in the lesson. That will be something I'll have to say at the beginning of next lesson.

   I said one thing I do care about is whether they arrive on time. There's only one lesson where they're coming from break which is the one they can have most control over, which is actually next lesson so I said that the lesson starts at 11.30 so they must make sure they're here at 11.30. And someone said 'What happens if we are late?' And I said you make up the time in break time or lunchtime. Quite an interesting question!

1.19 *LB:* That means they're intending to be late.

1.20 *AC:* Yes, exactly. So, we finished all that. Then I said, OK, a question I asked was 'How many names would be said by everyone? Has someone got an answer to that?' And someone said '636'. So, I said 'OK, how did you get that?' and he said '26 times 26'. So I wrote 26 times 26 on the board and it later transpired that he'd done that in his head so I'm not sure that he'd got the sum right but he was certainly attempting to do it in his head which was interesting.

1.21 *LB:* Numeracy strategy?

1.22 *AC:* Yes, absolutely. And so I said, 'Has anyone got any comments on that? Anyone think that that's right or wrong?'

1.23 *LB:* Same or different? No right or wrong you tell 'em and then you ask them if it's right or wrong [smiling]. Oh well.

1.24 *AC:* Did I say right or wrong? I can't remember. I might have said does anyone agree or disagree. And someone said I'm not sure that's right because the first person only said one name or he said not everyone said everyone's name for the 26 times 26. So, I again asked for any other comments.

1.25 *LB:* So, 26 times 26 that's if everybody says everybody's name and that's why you get the sense that it's not right or wrong because you have to also be in that space so you're not asking questions for which there's only one answer. What you're trying to do is to understand the space and where's 26 times 26 coming from?

1.26 *AC:* So, these guys said you've got to do 1 add 2. So, I then asked for any comments about that and then I've lost some of the detail around here. I think I then went back to the original person and asked him what he thought about the second suggestion and I think he said 'Yes, I can see that that needs to be 1 plus 2 plus 3 all the way up to 26'. So, I wrote that. I know there were a few other comments then at that stage and then I said 'OK, well the question was 'Can you work out how many names were said?' so I'll give you a couple of minutes to try and see if you can work on that and maybe see if you can work on a quick way of doing that.'

   So some of them said could they use a calculator to which I said no. And so some of them wrote up 26 times 21.

1.27 *LB:* 26 times 21?

1.28 *AC:* 26 down to 1 and added them up. Some of them were doing things like adding the first four up and then doubling that and so after a few minutes I stopped them and I asked for what different answers people had got and so we got a list of about 12 different answers.

   I just wrote them up at that stage and then I asked if . . . the reason . . . can anyone give me a reason behind any one of their answers. Some of them said that they'd just added them up but one girl said 'Well, I added up the first four and that came to 10 and then I thought that I needed 5 lots of . . . oh, no, that's not going to work'.

   And she said 'Oh, dear' and I said 'That's a really nice example of how working together mathematically as a group that in talking about how you're doing something you can recognise where you've gone wrong.' So that was nice.

1.29 *LB:* You're definitely right and wrong aren't you. The fact that you said it's alright to make mistakes, that had been last year's kids' way of describing it and what you're then giving them is that maths is still about right and wrong. And I actually don't think mathematics is about right and wrong, I think it's about explaining why you're doing things and doing methods and actually trying to do something in this way like the 26 times 26, you can actually go from there in fact to the 1 to 26. So there isn't that sense of it being right and wrong and as long as I'm aware of that which I'm doing that's the important thing which is why all this talking's important. But that would be a metacomment, about yeah, it takes our thinking on to articulate. That's a good example, and mathematician's do it all the time.

1.30   *AC:* That was Sophie who said that and she was still actually thinking about her things – she was still actually hooked in to something like that even towards the end of the lesson I think. Then rather mysteriously out of somewhere someone said well I did 26, the guy who'd done 26 times 26 he said well I did 26 shared by 2 to give me 13 and then I did 13 times 26. So, I said, 'OK, can you explain why you did that?' and he said 'Well, some people said all the names and some people didn't say all the names and so it's sort of 13 times 26'. I thought this is strange.

1.31   *LB:* He's basically averaged it.

1.32   *AC:* Absolutely. It's lovely. So I said, that's fascinating, I said, OK, could someone else, not Liam, say what they've understood him to mean by that? And someone else didn't quite say average but sort of said again in their own words this sense of some people saying more, some people saying less. And so I wrote out 1 plus 2 plus 3 plus 4 and I said so what would you do on this one? And he said, well I'd do half of 4 is 2 and then 2 times 4 to give us 8. Then someone said, it adds up to 10, 1 plus 2 plus 3 plus 4. So, I can't remember whether it was me or someone else then said 'So, if we look back to 1 plus 2 plus 3 plus 4 what is the average of these ones?' Then someone said 'It's 2 and a half'. And so there was a sense of if you do 2 and a half times 4 it does give us 10. Then some girl said 'We need to do 12 and a half times 26.' And so then I looked at them both and I think it was me who then said, but the 2 went to 2 and a half so it could be 13 and a half times 26. So, I asked someone to work that one out and I said you could use a calculator for that one. So that came up with an answer.

It still wasn't one of the ones that anyone else had got even though some had done it. Yes, it was a new one. So then I said 'could we test if this works for a smaller one?' So we agreed to do 1 plus 2 plus 3 plus 4 plus 5 which would then be three times five which was 15 which did seem to work. Which did give us the same answer. So, at that stage it was about 10 minutes to go or something so I said that that seems like a really nice stage and so what I'd like you to do for homework is to write up what this problem was that we've been looking at and write up what you've done to think about it and carry on trying to work at it and we'll talk next time.

1.33 *LB:* That's quite nice.

1.34 *AC:* I was a bit unhappy about it because I recognised that it was the first time they were going to write in their maths book was a homework which felt that it might be inhibiting for some of them. I was aware that I would have liked them to have done some writing before. Maybe getting them at that stage to have written out what the problem was.

1.35 *LB:* Maybe have written out the problem and some questions that they were thinking about. So, what happened then? How did you get rid of them?

1.36 *AC:* Then, well we just had a bit of time left so I'd wanted to do some chanting and some G chart work in the first lesson so they all wrote down their homework and I got them to pack that away then I said 'Something that we'll often do is do some oral work at the end which you may be used to' and I said 'quite often we'll be using this chart where I might ask you to do something and I'll point to something and you'll have to chant that response'. So I did some multiplying by 10, so I think we started with 6 or something and I said 'if I pointed to 6 what would you say?' and they said '60' so I said '6' and some said '60' and we did that three times until it was up at the volume which felt appropriate so we did that and quite quickly we got to 0.6 and 0.03 and things and they were all able to do that, they were up for that.

0.1	0.2	0.3	0.4	0.5	0.6	0.7	0.8	0.9
1	2	3	4	5	6	7	8	9
10	20	30	40	50	60	70	80	90
100	200	300	400	500	600	700	800	900

1.37 *LB:* So they knew it was looking down? It was immediate?

1.38 *AC:* Well, yes, it was pretty much immediate. The 0.03 there was a stammer for the first one, the first time I did it but then after that it was pretty much there. So we went right down to, we went off the chart, 900,000 went to 9 million and then I asked someone how they were doing it and they said well I just go down one on the chart. So I said OK, so how do you think you'd multiply by 100 and he said well I just go down 2. So we did some work on multiplying by 100. I didn't do dividing by 100. I don't know why because that would have been good to do as well.

1.39 *LB:* You've got a year.

1.40 *AC:* Exactly. So then I did some complements to 10 after that and then complements to 100. And complements to 10, again, the one boy who'd been the worst on the names seemed to be the one person who didn't seem able to do that. So we did the 100s so they all went 'intake of breath' but then they could do it.

1.41 *LB:* You had that! [surprised]

1.42 *AC:* I said you're all so good at the 10s we'll try 100. And they were all 'Oh, no' but then it was easy (60 – 40) and I said, so why is it so easy? They said it's just the same, you add a nought on the end and someone said, 'you can do 1000'.

1.43 *LB:* So what is the metacomment?

1.44 *AC:* I suppose extending the pattern.

1.45 *LB:* This is exactly what mathematicians do, they'll extend the pattern, we'll all get used to that and expect there to be structure. There's this sense that 'how do the kids know that they're not randomly doing a 1000?' They can't know that without the metacomment.

1.46 *AC:* Then I tried a few 650s and 750s but that was getting then quite hard for them.

1.47 *LB:* And that takes time. But it sounds as though they can do those things immediately, the 60, 40 even.

1.48 *AC:* No, I was really impressed.

1.49 *LB:* And if they actually have that sense of 'intake of breath' then it means that even in their first lesson they have the sense of learning something. Numbers that they thought were big, because primary schools I'm sure don't do big numbers and I don't see why not. It's just the same. Wow!

1.50 *AC:* So that was it and I let them out table by table. Then one person stayed and two people were being quite silly at the end, so I spoke to them at the end even though it was the end of the lesson. And one boy when I asked them to get their homework diaries out said 'I haven't got mine' and I could see it in his bag. So, I told him to get it out and I actually phoned his mum and said 'Very surprised in the first lesson, something I can expect hardened year 9s to do when they don't want to write down a homework'. She said she'd have a word with him.

Transcribed: 12/9/1999 Edited for book: 06/02/2001

APPENDIX 2

19/9/95 – Evening in London

2.1 *LB:* OK. I want you to think back over the first lessons you took this week and to choose one group, I don't mind which one and what I'm going to ask you to do is to try to re-enter that room, try to remember what you were thinking, what you were doing, where you were going as you walked through the door and then just to start telling the story of the lesson and from time to time I might ask for more detail or we might get into a conversation but for the beginning just tell the story and I'll take responsibility for that. Have got a lesson? So, say which class and how many kids.

2.2 *AC:* The year 7. There are 28 of them. And it was quite strange because it was my first lesson with them and it was in the computer room which has a big central area of desks which can fit about 15 people and then computers around the outside of the room where the rest of them sat. I wasn't very happy being in there for the first lesson because I didn't particularly want to get onto . . . well I wasn't sure that . . . there was this possibility . . . I was meant to be doing some computing module with them but I wasn't particularly keen on doing that for the first lesson. So I came in slightly undecided and feeling I might have to do this computer thing but not being particularly happy about it.

2.3 *LB:* Can you try to stay with the story of the lesson and try at the moment to hold off accounting for rather than account of?

2.4 *AC:* Basically I was walking in not quite sure exactly what I wanted to do with them. I came in. We have these rules for mathematics lessons that I've never given to a class. We're meant to do that first of all. So I decided that I would with this class. And that actually felt alright 'cos it just got them quietly working on something [copying rules into their books].

2.5 *LB:* Had you met this group before?

2.6 *AC:* No, it's the first time I'd met them.

2.7 *LB:* So, you walked through the door and there's – where are they? Are they outside?

2.8 *AC:* Right, OK. So they're all waiting outside this room and I tell them to fill up the central area of desks first and they do that and then they start having to sit next to the computers on the desks by the computers.

2.9   *LB:* Had you talked to them at all outside the room?

2.10   *AC:* No, I don't think so. And then I handed out exercise books and names, tutor groups that kind of thing. Having introduced myself...

2.11   *LB:* What do you do?

2.12   *AC:* I just said 'My name is Mr Coles'. Actually with my tutor group I said my name is Alf Coles but I said unfortunately we're not meant to use that name in addressing me. But with this group I just said my name is Mr Coles and I will be taking you next year. I quite liked with the tutor group saying my first name 'cos then it gets that issue out of the way which inevitably comes up. I think they were all excited about being amongst all these computers and there was a sense of expectancy. There was a sense of energy in the room perhaps about what was going to happen. But the first thing I did was write up these seven or eight rules, I don't know what they were all about, drawing margins and writing in blue and black ink which were all quite strange, quite strange for me, but I quite liked it just got them working and it settled them down and gave them something to do.

2.13   *LB:* So, you were writing things down on the board so that they would write them down in their books.

2.14   *AC:* In their books, copy them into the back of their books. And there were obviously people going at very different paces but there didn't seem to be anyone who was completely lost. They all were able to kind of write.

2.15   *LB:* So you were writing this on the board.

2.16   *AC:* And circulating a little. It's quite difficult to circulate effectively in that room.

2.17   *LB:* And how much of you being in that room affected you doing the rules?

2.18   *AC:* I think probably quite a lot. I think I'd have been far more inclined to just get on with some maths if I'd been in my own room. Yes, it was probably partly not particularly wanting to be in that room anyway. The sense of actually taking up a bit of time so I wouldn't have to get onto this sort of computing module in the first lesson. That definitely was part of the decision.

2.19   *LB:* And what are they doing?

2.20   *AC:* They're pretty silent. Writing down from the board. There were murmurings.

2.21 *LB:* So, can you remember what you were doing, was there any intention?

2.22 *AC:* Yes, to try to see how far everyone had got up to. I sort of wrote about four or five rules and then went and had a look at what everyone was doing. There was this thing about the board, the light, a bit of a palaver about the blinds, the lights, I'm not sure why. There were some questions at that stage. And then as most people began to finish that I decided to play a name game which was the one thing I'd decided I was going to do. I think that was the one thing I was sure I was going to do whatever happened in the lesson where the first person says their name and the second person says 'You're Ryan, I'm Tom' and the third person says 'You're Ryan, you're Tom and I'm Richard'.

2.23 *LB:* Before we get into that how many of the rules of your classroom do you remember? I have a question mark about 'what expectations do you share?''.

2.24 *AC:* Probably none. Well the rules were; 'must draw margins' and 'underline date for a piece of work' and 'write in black or blue ink', 'mustn't use Tippex', 'homework must be attempted on the day it's set', 'pencils for diagrams' and 'you need to bring pencils and rulers and calculators'. That's about it. One or two others. But that was the gist of it. They weren't so much the expectations, I mean I probably said something about value I placed on silence and said things about it being important that people don't talk when I or someone else is talking. But that wasn't one of the things they were writing down.

2.25 *LB:* I suppose one of the different things is that you are a new person to the class, so there is something about them looking at you and you looking at them and it's a time when some would say you get more attention then any other and first impressions of people stick so if there's something in that first impression that you really want to say, you'll always come with that first image. But then it depends on you being clear about what it is that you want to say.

2.26 *AC:* It's something I don't feel I've resolved anyway. The rules in general. That's where the doubt comes in in the first lesson, that I'm not really sure overall exactly what I want to expect.

2.27 *LB:* It just takes time. So, anyway the point which you'd reached is the beginning of what you call the name game. Try and go back into that.

2.28 *AC:* I think I started off by just asking the first boy on the left what his name was. And so the second boy to his right said his name and your name and the third boy and by that time people had twigged and there were these sort of gasps from the girls who were at the end of the line. And I was slightly concerned 'cos there are a few statemented pupils and I'd played the game before in my own tutor group and had been able to pick out one of the weakest boys to go first and I was slightly concerned that I might have got one of these very weak people right at the end which would have been quite difficult for him. But in fact it seemed to be going quite smoothly and I was prompting a little but basically people were ... yes it was actually a week into term and they'd been spending that week as a tutor group so, yes, that's right, so I wasn't sure if I was going to play in fact because I wasn't ... my first question was 'Do you all know each other's names?' and it was only because I got quite a few 'Nos' that I started to play it. So, there wasn't even that that I was sure about when I went in.

2.29 *LB:* So, one message you give by asking that question and then responding to it is immediately that you are not going to be the sole authority in the classroom. And that there are going to be some questions you ask of them that define your actions. Whether you're aware of that being a message or not, that would be a very strong message.

2.30 *AC:* That wasn't conscious at all.

2.31 *LB:* Because it's the fact that they said 'No' that meant that you did play the game and in a sense they would probably pick that up.

2.32 *AC:* That wasn't a conscious thing at all.

2.33 *LB:* Right, so they are certainly picking up that what they say is valued and that it makes sir do other things. OK, then what?

2.34 *AC:* Then we got about half way around the class and one of the things when I've played the game before that I feel hasn't been very good for people who've been early on is they tend to switch off. Not always, but often because their name is being said they are often attending but there seems to be a blank period in the lesson for those people perhaps. I said if you've said your name you might like to think about how many names we're going to say in total. I didn't sort of repeat it, I just said that once.

2.35 *LB:* One feeling would be, if the task is to learn names then the ones who say the names first are still going to learn them and the way that energises them is that at some stage when it looks as if they've turned off, I just look at them and say 'So, you think we're only going around once?' Having found that one I give you that one for free.

2.36 *AC:* That's brilliant. I'll use that. I hadn't pushed this maths bit. I'd literally only said it once. I just said some of you might like to think about how many names we're going to say in total. That was it and then we continued. Then as I finished off saying all the names, and they were all attending to that, that was good. Then some girl had worked it out on a calculator and some girl, I think she just shouted out the answer of what she thought they all added up to, so then I asked her how she had done that and she just said she'd added up 1 plus 2 plus 3 plus 4 ... yes, no, it was funny because they'd got the answer already. They'd worked out that there were 29 of us, or I'd said that there were 29 of us including me, and they'd worked out how many names there were and I said at one stage that I didn't say my own name so everyone said ' Say your name' to make it square with their answer, which was quite funny, so I dutifully said my name, which perhaps wasn't the right thing to do, I could perhaps have got something out of that, but ...

2.37 *LB:* Makes for less tension if you just do what they say. So, what had they done on their calculator?

2.38 *AC:* I think they'd just done 1 plus 2 plus 3 plus 4 ... asking afterwards, to the whole class that's what she'd said she'd done. So then I sort of wrote up the first five in the sequence and I asked if there was an easier way we could have started adding these numbers, perhaps we could have started pairing them. I think I mentioned the pairing idea. And someone suggested we could add the first and last and so I drew a few groups and arrows on the board and asked everyone to try and explore how many pairs for 30 or I might not have been so explicit, suggested look at this way of counting.

2.39 *LB:* Have you any sense of the different ways in which the rule for that situation might be got?

2.40 *AC:* There's writing the sequence once and then writing it backwards which is quite similar perhaps to the pairs, and then there's the pairing idea. I remember even at that stage I wasn't sure that I was going to continue with it. In the end this occupied the rest of the lesson but I think the reason I did is that someone was quite vocal about how this might be done and I think, I had been quite explicit about this other way of adding and some of them had picked up on that and so I'd gone with it. I think why I continued it with this suggestion from the pupils is that I didn't want to drag them straight to it. I was quite prepared to abandon it at that stage but because someone had seen something I was prepared to go with that, but I didn't really open things out for other ways of looking at it. In the end everyone, certainly for 29, was looking at the pairing idea.

2.41 *LB:* There's a very strong visual image about handshakes which is different to something like the name game and it's basically the same problem. The handshakes problem, that forces something, people to actually generalise straight away and the 'How many times has the name been said?' very often goes the way you've described adding general numbers together and it fascinates me that the context can change and our access to the rule is in a different sort of way. The fact of the shaking makes you almost draw a diagram and in the end the minus one divided by 2 is just somehow, is seeable. Just being able to describe it in words. And I don't get the same sense of this one being accessible in that way.

2.42 *AC:* Yes, I suppose, I did have a sense that everyone was just going to add up 1, 2, 3, 4, 5.

2.43 *LB:* OK, so, how did we get from you working on the board to ... what was happening at that stage? You did the first five, the last five ...

2.44 *AC:* Then there were a few suggestions about from the class about how we might do the sum more quickly. Then I think I suggested that everyone tried to do a quick way of finding the answer. And for a lot of them they hadn't found the answer. There were only a handful perhaps that had gone through it. So, for most of them it was trying to find the answer at all, and for the ones who'd done it it was trying to find a more efficient way of doing it.

2.45 *LB:* Did any of them try to do the simpler cases bit?

2.46 *AC:* No. They didn't. My suggestion to the ones who'd got there quickly was to ... I wasn't sure whether it would have been easier to say 1 or 2 or jump up to 50. Both those things were going around. It's much easier to see if you go up to 50 straight away. People were forgetting that the pairs would be different, rather than 1 and 29, they'd be pairing still to make up to 30. Perhaps the jump to 50 you can much better do that. There was a comment from one girl after she'd done the problem for this class, I suggested that one thing they might work on, she said 'You're just trying to make it more complicated for us' or something, I should have spent a bit more time to explain what I was doing. That was one of the things that stayed with me after the lesson that I wish I'd spent a bit more time answering that question properly, tell her what I was doing. But there were various interesting suggestions because with the 29 we had this 15 left over in the middle that couldn't be paired and some people were happy saying there were 14 and a half pairs for 30 which felt interesting, but there were lots of different ways of dealing with this 15. And then going up to 50 where there isn't a singleton left in the middle, I don't think, that again caused some interesting differences. I didn't really bring them back together again but looking back, that could have been quite nice because I offered to them to write it up that evening, but one girl at the end had kind of found a general formula for how many people there were in the class and a few people were getting towards that and there was one girl I spent a bit of time with who'd got 29 plus 1, 28 plus 2 and gone all the way down to 1 plus 29 and so I asked her about this and she realised that she didn't need half of them. By the end she'd just about, I'm not sure she did get it absolutely right, but she'd kind of got it. But there was quite a wide difference.

2.47 *LB:* The question there would be 'Is that something you're trying to set up?' so it's back to this 'what do you want to happen in that kind of lesson?' which might set a pattern for what might happen in other lessons. Can you remember what you told them about what they had to write?

2.48 *AC:* I suppose a lot of people had copied, 'cos I'd written 1 to 5 and 25 to 30 on the board and then drawn circles and arrows between them and a lot of them had copied that or wrote down something very similar but writing out the whole sequence, so I suppose that was the commonest way of writing out the sum. I was probably encouraging them to be writing down what they were doing as well. It certainly wasn't everyone who was doing that.

2.49 *LB:* Can you just try and think of a different first lesson with a different group and give me a sense of how that might be different in terms of what you might have been trying to do or how it might be the same?

2.50 *AC:* With my year 12 the first thing I asked them was to write down an equation that they could solve and one that they couldn't and then to write down the hardest one they could and the easiest one they couldn't and again wanting to get across, I don't know, I suppose at one level wanting them to get in touch with what they could and couldn't do but also trying to get across some sense of autonomous working.

2.51 *LB:* That's also about choice in that you're going to adapt to their information. One of the hard things about working like that is once you've got the information you've got to decide what to do with it.

2.52 *LB:* I've been trying to put detail in. Bernard Murphy's piece of writing in *MT131*. He's a local head of department. He's the one who the classes three years on are behaving completely autonomously and I don't know what to say to the kids in his room. I'm not usually dumb-founded, they're in a completely different space to one I'm used to being in. He has been doing writing in detail. The trouble with detail is it's fractilic in the sense that you can never get enough of it, because any individual person reading it still won't know what to do. But I do think it's important, to focus more on teaching than on learning. This is unusual. In the end it's what the kids do. My heart's in that place when I'm in the classroom it's just that that can't be freed unless the teachers find out what they're doing. So, what would reading the detail give you, or what do you think reading the detail would give you? It's maybe more that you can't turn it? I don't get a sense that you can turn things into, you can't see something that you're reading in *MT* and it's become yours – would that figure?

2.53 *AC:* Myself? Or?

2.54 *LB:* In a sense I wouldn't necessarily miss the detail in some *MT* articles because in reading the idea, like arithmogons, it would become something and that, I was trying to give you a sense of that and maybe there's this sense in which you want more detail because you're still at the stage where, in a sense, that might be helpful because it gives you some more to try.

2.55 *AC:* It was just exactly, it was all the things that perhaps these people had just got their classes doing anyway, all the things like, discuss in pairs and bring it back to the class and I'm thinking well, how on earth do you do that?

2.56 *LB:* Well that's in fact the test of writing in detail is to try to open up some strategies where people will have things to do given a whole range of, it's this three layer thing, given a whole range of different central strategies, ideological positions and articulated purposes, a sense of micro-strategies below that? and I had a conversation where a teacher said something like 'Oh, that means you'll be able to tell me the best way of doing something' and I said it's more complicated than that, it would depend where you're coming from because everyone would react differently. There has to be a sense in which there are some things you could say. I've got a strong sense in working with you that that's dangerous where, I know that these things are transferable so why in the hell don't I, instead of writing this very complicated, convoluted thing, why not go ahead and do it. There's where I don't believe that it's true, that it is possible to just transfer.

2.57 *AC:* You don't believe it's true but you know it's true?

2.58 *LB:* Yes, well, for particular people. And that's the question mark about working with you in terms of this because if the transfer is, in a sense, too direct then there's no space for you in the middle, and yet, the challenge would be, I'm going to have to think about it for a bit and then come back, that this thing about you developing and tracking that development and whatever might be this image and it's multiple images, it's not just one, where in a sense you do find your voice and you do see a vision about how it's going to be and part of that is picking up from the kids.

2.59 *AC:* Yes, absolutely. I got a real sense of how I could work in the class today. How I could get towards working in the kind of way that I want.

2.60 *LB:* So, say, try and say the sense of what this might be.

2.61 *AC:* A sense of what it's like to create a space for students to work on their own images. A sense of how that could be maintained. Picking up from what students have said at the beginning of class – the thing that you said on the first day that you only hear what, things that you're open to and that there isn't a sense that by some people giving their method or explaining what they're doing it's not that you're taking away the journey from the others it's helping them along.

2.62 *LB:* If the message that you're explicit about giving that you're only cheating yourself if you don't understand then it's not cheating to copy it's not cheating to talk to your friends it's not cheating – the only cheating yourself is if you don't understand. And some classrooms are very explicit about that so, another description of you at the moment is that you have no metacommenting. Some teachers when they are teaching are metacommenting. Some people think of this as patter - anything from margins to 'remember to reflect on what you're doing aswell as do it.' That's quite striking, I don't think I've spotted any. Again, it's not good or bad. I think what happens to teachers is that they become very clear about what they're trying to do, so all those metacomments become completely standardised and in a sense they become close to their behaviours – it's a sort of autopilot. Right down to a kid comes up and says 'I've got a problem with this' to 'OK, you tell me'. It's a metacomment because it's saying 'I'm not going to engage with this until you tell me'. Sometimes that's enough.

2.63 *AC:* One of the things that has come out in some of the early lessons is when I've been quite explicit about 'I'm not really here to give answers.' That came out in the year 12 class [aged 16-17] at one stage. They were saying 'Is this right? Is this right?' and I kind of resisted.

2.64 *LB:* In fact that's not around. That would be an example. That is one. In a sense an awareness of that and the behaviours developed.

2.65 *AC:* With my year 9 class who I had in year 8 one of the things that I said to them at the beginning is that 'one of my aims this year is to try to make you independent thinkers', although who knows what independent means. That felt quite nice because I could come back to that.

2.66 *LB:* That was in a first lesson.

2.67 *AC:* Yes. That gave some power to when they said 'Can you tell us how to do this?' that I could say 'I'm wanting you to do this independently.'

Transcribed finally: 21/12/99 Edited for book: 06/02/2001

9 REFERENCES

[1] Data taken from: 'Developing algebraic activity in a 'community of inquirers'' Economic and Social Research Council (ESRC) project reference R000223044, Laurinda Brown, Rosamund Sutherland, Jan Winter, Alf Coles.
Contact: Laurinda.Brown@bris.ac.uk or Laurinda Brown, University of Bristol, Graduate School of Education, 35 Berkeley Square, Bristol BS8 1JA, UK

Aristotle (1980) *The Nicomachean Ethics*, Oxford: Oxford University Press

Banwell, C., Saunders, K. and Tahta, D. (1972) *Starting points*, London: Oxford University Press

Bateson, G. (1979) *Mind and nature*, New York: E P Dutton

Brown, L. (1995) 'The influence of teachers on children's image of mathematics', in Meira, L. and Carraher, D. (ed), *Proceedings of the Nineteenth International Conference for the Psychology of Mathematics Education*, Brazil, Recife, **2**, pp. 146-153

Brown, L. (1988) *The influence of teachers on children's image of mathematics*, unpublished MEd dissertation, University of Bristol, Graduate School of Education

Brown, L. and Coles, A. (1996) 'The story of silence: teacher as researcher, researcher as teacher', in Puig, L. and Gutiérrez, A. (eds), *Proceedings of the Twentieth Annual Conference of the International Group for the Psychology of Mathematics Education*, Valencia, Spain. **2**, pp. 145-152

Brown, L. and Coles, A. (1997) 'The story of Sarah: seeing the general in the particular?' in Pehkonen, E. (ed.), *Proceedings of the Twenty-first Annual Conference of the International Group for the Psychology of Mathematics Education*, Lahti, Finland, **2**, pp.113-120

Brown, L. and Coles, A. (1999) 'Needing to use algebra – a case study', in Zaslavsky, O. (ed.), *Proceedings of the Twenty-third Annual Conference of the International Group for the Psychology of Mathematics Education*, Haifa, Israel, **2**, pp. 153-160

Brown, L. and Coles, A. (2000), 'Same/different: a 'natural' way of learning mathematics' in Nakahara, T. and Koyama, M. (ed.), *Proceedings of the Twenty-fourth Annual Conference of the International Group for the Psychology of Mathematics Education*, Hiroshima, Japan, **2**, pp. 153-160

Brown, L. and Waddingham, J. (1982) *An addendum to Cockcroft*, Bristol: Resources for Learning Development Unit

Claxton, G. (1996) 'Implicit theories in learning', in Claxton, G., Atkinson, T., Osborn, M. and Wallace, M. (eds), *Liberating the Learner*, London: Routledge

Coles, A. (2000) *Teaching strategies for developing listening and hearing in two secondary mathematics classrooms in the UK*, unpublished MEd dissertation, University of Bristol, Graduate School of Education

Coles, A. (2001) 'Listening: a case study of teacher change', in Van den Heuvel-Panhuizen, M. (ed.), *Proceedings of the Twenty-fifth Annual Conference of the International Group for the Psychology of Mathematics Education*, Utrecht, Netherlands, Utrecht, **2**, pp. 281-288

Cockcroft, W. (1982) *Mathematics counts*, London: HMSO

Davis, B. (1996) *Teaching mathematics: toward a sound alternative*, New York & London: Garland Publishing Inc.

Gattegno, C. (1974) *The common sense of teaching mathematics*, New York: Educational Solutions

Gattegno, C. (1987) *The science of education: part 1 theoretical considerations*, New York: Educational Solutions

Gattegno, C. (1989) *Gattegno anthology*, Derby: ATM

Goldsworthy, A. (2000) *Time*, London: Thames and Hudson Ltd

Gratry, A. (1944) *Logic*, Illinois: Open Court Publishing Company

Holt, J. (1990) *How children fail*, London: Penguin Books

Jaworski, B. (1991) *Interpretations of a constuctivist philosophy in mathematics teaching*, unpublished PhD Thesis, Milton Keynes: Open University

Kieran, C. (1996) 'The changing face of school algebra', *Eighth International Congress on Mathematics Education*, Seville, Spain. http://www.math.uqam.ca/_kieran/art/algebra.html

Lakatos, I. (1976) *Proofs and refutations*, Cambridge: CUP

Little, W., Fowler, H. and Coulson, J. (1973) *The Shorter Oxford English Dictionary*, Oxford: Clarendon Press

Mason, J. (1996) *Personal enquiry: moving from concern towards research*, Hampshire: The Open University (Researching Mathematics Classrooms, ME822)

Nicol, C. (1999) 'Learning to teach mathematics: questioning, listening, and responding', *Educational Studies in Mathematics*, **37**, pp. 45-66

Rost, M. (1994) *Introducing listening*, London: Penguin Books

Sutherland, R. (1997) *Teaching and learning algebra pre-19*, London: RS/JMC

Thera, N. (1996) *The heart of Buddhist meditation*, Sri Lanka: Buddhist Publication Society

Varela, F., Thompson, E. and Rosch, E. (1995) *The embodied mind*, Cambridge, Massachusetts: The MIT Press

Wheeler, D. (1970) 'The role of the teacher', *Mathematics Teaching*, **50**, pp. 23-28

Wheeler, D. (1975) 'Humanising mathematical education', *Mathematics Teaching*, **71**, pp. 4-9

Printed in the United Kingdom
by Lightning Source UK Ltd.
132570UK00001B/75/P

9 781900 355599